Malignatech

Lyden Pendleton

INFINITY
PUBLISHING.COM

ISBN 978-0-7414-5258-0

Published by:

INFINITY PUBLISHING.COM

1094 New DeHaven Street, Suite 100
West Conshohocken, PA 19428-2713
Info@buybooksontheweb.com
www.buybooksontheweb.com
Toll-free (877) BUY BOOK
Local Phone (610) 941-9999
Fax (610) 941-9959

Printed in the United States of America

Published February 2013

Preface

Man's struggle began simply enough. A primitive man got tired of being soaked by rain and bruised by hail, so he found shelter in a cave. Next came better and better weapons to hunt with and kill enemies. Then fire to cook tough meat. Then gunpowder, then computers, then rockets to Mars.

That struggle, by its very nature, led man beyond just the need to survive Nature's brutality. Man came to believe he could actually dominate Nature with his weapons and machines. Now, as the 21st century begins, man's struggle to survive has become twisted into a dilemma of mythical proportions: how does man dominate a force far more powerful and wiser than he will ever be. A force of which he is but a small cog in a large cosmic system.

This book is dedicated to one critical question mankind must answer very soon: Does man have the intelligence to realize that his attempts to dominate Nature are actually destroying himself?

Introduction

THE LUDDITES WERE RIGHT!

Almost two hundred years ago, at the dawn of the Industrial Revolution, a prophetic cultural war began in Nottingham, England. Workers in stocking manufacturing plants became distressed when machines that did much of their work were installed. The year was 1811 and England's economy was still reeling from war. Job security was therefore a big concern. Without such modern protections as benevolent judges and unions, the workers took their most direct course of action: they destroyed the machinery. These destructive riots spread into other districts and extended over much of England for several more years. Severe legal measures and post-war prosperity finally ended the Luddite movement.

In the 200 years since then, anyone considered anti-technology has been branded with this crude term. The implication is that any such person is either out of touch with the modern world or just a complete idiot.

As I shall document in this book, however, the Luddites were way ahead of their time and far wiser than anyone realized. Technology, especially in the computer age, has obviously created more jobs than it has eliminated. Many lifesaving inventions have enriched our world, but the price we are now paying for this higher material standard of living has become intolerable. Malignatech, technology's dark side, is not only destroying us and our environment but the pace of this destruction is also accelerating.

This book has no hidden agenda. I am not promoting any particular religion, philosophy, or brand of politics. Much of the insane destruction modern man is reeking upon himself and Nature is permanent. We need to stop malignatech before it stops us.

malignatech, n. The dark side of technology. A subtle mental illness that begins with the naive belief that technology is *inherently* good. By attempting to elevate man above Nature, this disease isolates him from Nature as it conspires to steadily destroy our natural world. Malignatech is a combination of irrationality, lust for power, and material greed. It originated in America and is rapidly spreading throughout the industrialized world.

Chapter 1—SAY GOODBYE TO THE BIG 3 R's

Perhaps the most serious threat to America's future is malignatech's insidious ability to destroy the essence of our culture: rationality. This relatively modern way of thinking concentrates on working from the natural world and doing what is logical. If everyone thinks the world is flat, sail to the edge of it and see if you fall off. Magellan accepted that challenge and had a long, exciting voyage to recount. If birds can fly, why not man? Build a device with wings shaped like theirs and soon man is the fastest winged creature on Earth.

That rational thinking was working fairly well for us until nuclear weapons came upon the scene sixty years ago. Suddenly, the rules of warfare were warped completely out of shape. Now a single bomb could destroy an entire city and hundreds of thousands of lives. America was in way over its head. Our government had far more destructive power at its command than all the other nations in the world combined. We were invincible … for a short while. Then the Soviet Union made a few nukes and the Cold War began. Both countries built *thousands* of these bombs, enough to destroy every major city on the planet. Rationality took a serious blow to the head with this Mutually Assured Destruction concept. Appropriately dubbed M.A.D., this doomsday concept created a radically new and depressing sense of helplessness among postwar generations. I still have vivid memories of grammar school air raid drills back in the 1950's. Bells rang and we all dutifully marched out into Central Grammar's long main corridor. We sat down against the wall close together, then at the teacher's command we all leaned in one direction, with one arm covering our head. It was quite cozy and sure beat doing fractions, but I remember thinking at the time: how is leaning on each other going to protect us from getting fried if an atomic bomb goes off anywhere nearby?

Then the '60's arrived, along with the Pill, all sorts of weird drugs, wild rock music, and a general feeling of anything goes. If we're all gonna fry in a huge fireball any day, why not have some fun now? With the future looking so grim, lots of young people were desperate to escape reality. Rationality took another major blow to the head when the drug culture emerged. Frying your brain with an acid trip to avoid thinking about having your body fried in a nuclear war is not very rational. That type of warped logic became seductive. As machines make more and more of the decisions we used to have to ponder, we are steadily losing the power to think rationally.

Our main mistake here is assuming that all technology is good and helpful. This leads us to malaxiom #1:

> "Technology's powers will always be used for evil purposes to the fullest possible extent."

Smoke detectors save us from fires, obviously, so radar detectors must be doing some good protecting us from radar, right? How dare those state troopers try to arrest drivers who want to race around and endanger innocent lives, right? Wrong!! Radar detectors are a shining example of malignatech at its worst. Why do Americans tolerate an invention that encourages people to break the law?! Why do ads for these insane devices tout the "fact" that radar can somehow get misdirected so the police can mistakenly arrest you for speeding when it was really the car beside you? There is no logic to the invention or sales of radar detectors, yet every other car seems to have one. Would the public approve the sale of "teacher detectors" so pedophiles could attack kids on school playgrounds? Probably not. Which can do more damage, a pedophile or a car doing 80 miles an hour?

As materialism and laziness become more the norm, machines step in to satisfy our needs. Want a bigger and more shocking thrill? No problem. We have a zillion violent video games and movies. Half the shows on television revolve around dead bodies, past and present. Our culture has become so warped (punk rock music), so distrustful (lying presidents and corporate executives), and so jaded (gas guzzling SUV's) that rationality is rapidly becoming obsolete. I'll explore this sad and shocking trend in coming chapters, but here are a few glaring examples of the breakdown of rationality in America:

* Hurricane Katrina—the most expensive natural disaster in our country's history was essentially man-made. By erecting levees that were not strong enough to withstand a Category 4 storm, by not maintaining those levees, by allowing the barrier islands off the coast to be washed away, and by building houses in an area of New Orleans which was nine feet below sea level, the city set itself up for disaster. To then have city and state officials blaming Washington for its lack of assistance was pure scape-goating. Those officials conveniently forgot to mention that a 2003 study predicting a disaster of Katrina's size was largely ignored by Louisiana. In fairness to those officials, the federal response to this disaster was incredibly inept.

* Southern California Brush Fires—in 2003 this area was devastated by fires that burned 750,000 acres, destroyed almost 3,000 homes, and killed 20 people. Why do these people *constantly* build homes amid tinder dry brush in deathtrap canyons?! Hey, the fancy houses

are insured. Just wait until the smoke clears and rebuild, right?

* Eating Contests—probably the most disgusting example of malignatech is the International Federation of Competitive Eating. This so-called organization sanctions 150 "chow downs" a year, where amateur and professional gluttons can see who can jam the most hot dogs, pies, etc. down their throat in the shortest burst of time. 1 In an era where third world starvation is growing at a staggering rate, how can any group in its right mind promote sheer, idiotic gluttony?!! And Americans wonder why poor countries dislike us! This only one of the seven deadly sins we are cultivating as sports. Unbelievable!!

Our second big R in jeopardy is responsibility (no, my big 3 R's go way beyond readin', writin', and 'rithmetic). This formerly sacred word is rarely heard outside courtrooms today. Most Americans seem to have forgotten the meaning of responsibility. How many protest marchers have you ever seen marching with a sign that states "I DEMAND THE RESPONSIBILITY TO…"? I doubt that anyone has ever written such a sign. Everyone wants more rights, yet very few people want the responsibilities which come with those rights. In fact, in our great "free lunch"[1] culture few people even realize that rights *always* come with responsibilities. Failure to accept those obligations leads to losing the rights. Don't want to drive sober? Welcome to sobriety checkpoints along the highway. Your right to unimpeded travel just took a hit.

Knowing this delicate relationship, many groups (corporations, lawyers, and—especially—governments) are making a major effort to get you to forget your responsibilities. Television is a perfect example of this trend.

Worried about your impressionable young kids watching excessive sex or violence on the tube? No problem. Just sit back and relax; the V chip will do the parental supervising for you. Sitting down with your kids and discussing what is appropriate to watch is no longer necessary. Now you can program your television set to be a parent and control what programs your kids are allowed to watch. Kids, who usually know more about computers than their parents, will never be able to figure out a way around the V chip, will they? Who's kidding whom!? So when your child finds a way to watch the increasingly violent trash on television then decides he needs to play gangsta rapper and bring a gun to school, you can just blame that stupid V chip for failing its responsibility. When kids give up their responsibility to be law-abiding members of society, they get confronted with metal detectors at the school door, unannounced locker searches, and armed guards with drug-sniffing dogs roaming the halls. These are all major features of prisons.

How many institutions in America teach that extremely vital relationship between rights and responsibilities? Do you know of any? Once rationality becomes so warped that everyone can custom make their own version, responsibilities become merely a speed bump on the road to unlimited rights and fun. Here are some blatant examples of attempts to avoid serious responsibilities:

* Gay Marriage—seems like a trendy, very tolerant-type advance in human rights. Two men or two women love each other and want to commit to that sacred relationship known as marriage. Seems logical, until you stop and realize the main purpose of marriage; namely, to have children. Creating and raising the next generation of Americans is the most difficult and expensive challenge two adults (by necessity, of the opposite sex) could have. NO OTHER task is as important to society as

parenthood. It requires love, patience, caring, vigilance, planning, sacrifice, and lots of money. It consumes two or three decades of a couples' lives. Gay marriage conveniently sidesteps these responsibilities. Two members of the same sex cannot have children together. What, then, is the motivation for wanting to be married, instead of just living together as a couple? Gays and lesbians obviously want to legitimatize their lifestyle with an official title. That's understandable, *but* to expect the same rights a married couple has is not only illogical, it's extremely insulting to the institution of marriage. Should gay and lesbian couples have survivors' benefits for life insurance policies and Social Security? Should they have spousal rights so that only one member has to pay to join an expensive country club? The financial aspects of this controversy make gay and lesbian marriage look suspiciously like another quest for the great American "free lunch."

An ironic sidelight on this matter is that the gay marriage movement probably owes most of its existence to a fatal disease by the name of A.I.D.S. Back in those free love '60's and '70's the hallmark of urban gay sex life was its frantic promiscuity. Gay bath houses in cities like San Francisco were booming. One quote in a newspaper from that city had a gay guy explain it all very simply: "It's impossible to feel guilty for having a lot of sex partners." With the arrival of the A.I.D.S. epidemic in the '80's that attitude gradually changed to a more monogamous one. When the grim reaper is staring over your shoulder, it's considerably easier to act straight and yearn for the altar.

* Pedophile Priest Scandal—this incredible disgrace of the Catholic Church is perhaps the most shocking recent example of avoiding major responsibility. A surprising feature of this bizarre moral breakdown has been the *entire* world's overlooking one critical fact here. Estimates of how long priests have been molesting children vary, some of them going as high as the last fifty years. Pope John Paul's term of office, which began in 1978, falls completely within that time frame. In other words, these crimes were being committed during his entire term of office. Did he know about these crimes? There is no way he couldn't have known! Those hundreds of millions of dollars being paid out to the priests' victims would have been next to impossible *not* to notice. The bottom line here is that Pope John Paul II, as head of the Catholic Church, was responsible for the last 26 years of the pedophile scandal. Even when it blew up in his face, he refused to become directly involved, letting the American cardinals take the blame. Pope John Paul's punishment for allowing these disgusting crimes to continue for decades, disgracing his church and costing it over $1 *billion* in damage payments? A couple of glowing television biographies and being put on the fast track to sainthood by the Vatican!! Who says crime doesn't pay. 2

When rationality goes out the window, responsibility—whether personal or holding leaders to theirs—soon follows. This sordid priest scandal is frightening proof of that relationship. Once rationality and responsibility are out of the way, the stage is set for the coup de grace on your rights. Failing to hold our leaders accountable is an open invitation for them to trample those rights whenever they want.

This pathetic trend has been going on in America for over seven decades, ever since Franklin Roosevelt outlawed gold ownership in 1933. Since gold and silver were—and still are—the only constitutionally-approved money, Roosevelt essentially told the American people that they were no longer permitted to own money!! Like the flock of sheep they are, Americans went along with this incredible violation of their constitutional rights until 1974, when gold ownership was again "legalized" by our benevolent government. Your federal government's lust for your rights has continued to grow with every passing year.

How does technology's dark side, malignatech, figure into this socialistic trend? Does anyone realize that technology is rapidly becoming not just our master, but also our executioner?

In the following chapters I'll explore this radical thesis into some of its most dangerous areas. By destroying the Big 3 R's, malignatech threatens our happiness, our prosperity, and our very lives.

Chapter 2—ENTER THE SILENT KILLER

Malignatech has its roots in a powerful force discovered in the early 1900's. Stress was first discovered by an Austrian physician named Hans Selye who moved to America at the turn of the century. Although he had no formal training in psychology, Selye created the concept of stress by decades of brilliant research which he compiled into a ground-breaking book titled *The Stress of Life.* He admitted that explaining his concept of stress in a book was motivated greatly by his desire to share with others "the serene and elevating satisfaction which comes from understanding the inherent harmonious beauty of Nature." Here was a gifted laboratory scientist who knew the importance of being close to Nature. He understood how science has shaken our blind faith in traditional values, to the detriment of society.

Doctor Selye's most famous discovery was the General Adaptive Syndrome or G.A.S., a collection of symptoms of the human body responding to stress. Under intense stress the G.A.S. appears in three stages. First comes the *alarm reaction* in which the body is alerted to a threat by the release of hormones. Secondly, the body calms down and adapts to a stage of *resistance.* If the stress condition persists long enough, that resistance stage eventually dies out, leading to a third stage of *exhaustion.* If this persists, the body dies. Luckily for us, this outcome is the exception in most stress situations.

Selye's genius was to see that throughout the human body's extremely complicated chemical response to stress the basic sequence of reactions was predictable enough to be readily measured, and, therefore, studied. This discovery revolutionized the understanding and treatment of stress by giving it a physical dimension. Stress was recognized as an unavoidable part of life, with some types like joy and exhilaration actually being beneficial. Further research revealed that stress plays some role in the development of 80 percent of illnesses, including heart problems and cancer. It is estimated that the economic toll of U.S. stress-related illnesses exceeds $200 billion annually! How much vital healthcare could such wasted money buy? 1

In keeping with Selye's work, scientists are now concentrating on measuring the key stress hormone, cortisol, which is released by the adrenal glands located on top of our kidneys. Cortisol is a key hormone involved in major bodily functions, including cardiovascular and immunological function, blood pressure regulation, and metabolism of food.

Let's examine America's top two killers and see how malignatech-induced stress is involved in their rapid spread. Cardiovascular (heart) disease is this country's leading killer. One person in four suffers from some form of it, with over 869,000 dying from it in 2007. 2

As I have personally learned through painful experience, the heart is our most hard-working organ. The fact that it can work nonstop for a hundred years is almost beyond belief. Our heart's strenuous work leaves it vulnerable, however. A recent Mayo Clinic study of heart disease patients revealed that those with the highest stress levels had definitely higher rates of rehospitalization and reoccurrence of serious heart problems. In a recent report in the *European Heart Journal,* researchers studied nearly 14,000 men and women for 21 years. These researchers concluded that "chronic stress is an independent risk factor for cardiovascular disease, particularly fatal stroke."

So what's causing all this modern stress? A major factor is simple inactivity. It's an unnatural state for humans; we were designed for action. All of our hundreds of "labor-saving" devices area turning us into sedentary weaklings. Less exercise means less oxygen getting into the blood stream, less blood circulating, reduced muscle tone and mass, then the big danger of obesity.

Now that the majority of adult Americans (60 percent) are overweight, society's perception of this unhealthy condition is changing. This new majority is gradually trying to win acceptance of their condition. Celebrities admit they're overweight and proud of it. Baggy clothes and "plus size," "full figure" fashions are everywhere. This gradual acceptance has been aided by the misconception that excess weight is more of a beauty problem than anything else. Health problems here are slow in developing, but far more serious than most people realize. The list includes fatigue, decreased circulation, mental fuzziness, sleep apnea (brief breathing stoppage), joint problems, general breathing problems, bone fractures, vision problems, hormonal imbalance, diabetes, angina (heart pain), stroke, and heart attack. 3

You would think the over 5,000 fat-free or low-fat food products on the American market should have put a clamp on the rise of obesity. In fact, just the opposite has happened! People believe these foods are safe to chow down on, not realizing "fat free" often means high in sugar and calories. The result has been a doubling of obesity in the last 20 years. Don't all the dozens of diets out there help? No! Their success rate of people keeping weight off for at least a year is about 3 to 5 percent. Why? Because diets fool the body into losing weight only until the body's natural survival instinct takes over and it starts storing fat for lean times. The human body does this in a surprisingly overly-efficient way, by increasing not only the number of fat cells but also the size of them. 4

Perhaps the most astounding flaw in our overall diets is the shocking amount of sugar we consume. We've been conditioned since early childhood to associate sugar in the form of candy with holidays and happiness. There are candies for Christmas, Valentine's Day, Easter, Halloween, birthday parties, carnivals, you name it. Feeling depressed? Have some sinfully delicious chocolate fudge! The fact that modern refined sugar has no food value is bad enough. Add in the fact that sugar in the body becomes acidic and the result is severely weakened immune systems and a national diabetes epidemic. Eating excessive sugar can also weaken the pancreas, the organ which produces the insulin which allows your cells to burn sugar to give you energy. Also, high insulin levels that can result when a healthy pancreas gets hit with excessive sugar create bad cholesterol, high blood pressure, and thicker blood. The end result is a weakened heart and clogged arteries. Numbers confirm this scenario. Two-thirds of heart attack victims have blood-sugar problems. 5

Three hundred years ago the average person consumed approximately one pound of sugar annually. Now that amount has jumped to 130 pounds a year. 6 Let's examine sugar using malignatech's three components. First, most of us know that sugar is not good for our systems. It causes tooth decay and diabetes. But it tastes so good, so we pig out! *Irrationality* sets in big time here. Sugar is a cheap ingredient that can make ordinary food like soda, candy, and pastry very addictive. The more customers companies can make addicted to their products, the more power these companies exert over us. Chocoholics will pay big bucks for fancy brands. How about *greed?* Turning a nation away from natural sweets like fruit toward manmade sweets has been very profitable for our food industry. Malignatech and sugar, a perfect match.

In summary, heart disease in the last century has gone from being almost unknown to a national epidemic. How has this happened with all the advances in medicine since 1900?

Sedentary lifestyles, obesity, poor nutrition, pollution, toxic medicines, and violent entertainment have all combined to assault our bodies with an assortment of stresses that have hammered our dutiful hearts.

America's number two killer, cancer, is scarier than heart disease. Although cancer kills slightly fewer of us—500,000 annually—it has a Grim Reaper-like reputation of pouncing on healthy Americans without warning, appearing to be a ticking time bomb hidden deep inside each of our bodies. Sadly enough, that's exactly what cancer is, an aberration of normal cell division which creates a tumor. As with heart disease, cancer was essentially *unknown* a century ago, when only 1 percent of us were stricken with it. Today 50 percent of us will get some form of this disease. 7 Those are very scary odds.

I could obviously write a whole book about the many types of cancer, their causes and cures. For this discussion let's focus on the likely causes and their relationship to technology. Since most of our genes have probably not turned self-destructive in the last century, we can rule out genetic tendencies as a major cause. That leaves toxins from our environment or our diet as the probable cause. Over half of all cancers have been linked to lifestyle factors, the chief one being nutrition.

Enter the malignatech connection. Beginning in the 1940's American farming veered away from growing organically to the widespread use of manmade fertilizers, pesticides, and antibiotics. Processed food was born and our bodies have never been the same. Lack of fiber here slows down the organs of elimination, leaving our bodies in state of chronic toxicity. Our modem processed food diet harms us in four critical ways:

Toxic Fertilizer

Replacing organic fertilizer such as manure with chemical fertilizer was bad enough. You won't believe what fertilizer companies are putting in their product today!!

Ever wonder where all the *thousands* of tons of toxic industrial waste are disposed of? Ever drive by a dump and see a sign that reads "INDUSTRIAL WASTE DUMP"? You very likely haven't. Then where do all those freight cars of waste with all their heavy metals disappear to? They are mixed into the fertilizer that is put on the crops that grow into the food YOU and YOUR FAMILY eat!! Impossible, you say? No, it's standard practice. Single companies dispose of millions of pounds of dioxins, toxic metals such as cadmium, beryllium, chromium, lead, mercury, and aluminum. Also, arsenic and uranium!

How could this happen in green America? In 1978 the Environmental Protection Agency (EPA) delegated the control of fertilizer ingredients to the states. Since then over half the states have allowed hazardous waste to be put into fertilizer, instead of waste landfills where it belongs. This is done by a trick that would make George Orwell "green" with envy. By declaring the hazardous waste a product—essentially an ingredient—a manufacturer can put as much waste into a fertilizer as it wants to!! *Now* do you believe in malignatech?! The presence of these wastes is not required to be mentioned on the fertilizer bags, so no one knows this toxic secret except the states and your EPA, both of whom look the other way. After all, it *is* recycling.

Here's the really scary part. All those millions of pounds of toxic heavy metals stay in the soil for *centuries]* 8 We have allowed our soil to be permanently contaminated, all because few people care about genuine nutrition anymore.

Nutrient Depletion

Processed foods are high in substances like fats, sodium, and sugar which make us feel full but crowd out vital protective nutrients like potassium, selenium, zinc, chromium, and magnesium. As shiny and colorful as they look, fruits and vegetables found in supermarkets are hybrids that are shadows of their former selves. University of Texas researchers have discovered that levels of calcium, vitamin C, riboflavin, iron, potassium, and protein in fruits and vegetables have declined considerably since 1950. As you will recall, that was just after the introduction of chemical fertilizer. Food today is considered by commercial interests to be a commodity to be made as profitable as possible. Plants are bred for appearance, resistance to spoilage, and water content. The crowning insult to consumers is the practice of picking fruit green before it has a chance to ripen.. Without that extra sun, fruits can't develop sun-related nutrients, which give us disease protection. Cherries picked green have half the vitamin C content of ripe cherries. Spraying produce with ripening agents won't bring back those lost nutrients. 9

Free Radical Damage

This relatively new theory of a major cause of cancer and aging is extremely complicated, so I'll simplify the discussion as much as possible. In the late 1960's a Shell Oil research chemist named Dr. Denham Harzman was working in a new field called oxygen chemistry. He observed that oxygen reactions had a tendency to age rubber, and theorized those same oxygen reactions could harm human organs and contribute to aging. Dr. Harzman's theory got attention from other researchers, and the free radical theory of aging was born.. It proposes that our body's normal metabolism and exposure to environmental toxins produces unstable agents called free radicals. These reactive agents damage cells throughout our body. We have an internal protection against this process in the form of antioxidants. Unfortunately, levels of these vital substances decline with age, creating an imbalance known as oxidative stress, which has been associated with aging diseases such as heart disease, cancer, and Alzheimer's. 10 Once again, we're hit with a form of stress.

While many in the medical field are still skeptical of the value of taking antioxidants to help fight free radical damage, a growing amount of research validates this trend of anti-oxidant foods and supplements. Iron and copper, consumed in excess, have been shown to promote free radical damage. Interestingly, certain antioxidant foods have been shown to attack some free radical substances and not others. For example, beta-carotene stops singlet oxygen but not hydrogen peroxide. 11 An Italian study done in 2000 found that a group of centenarians (age 100 or older) had significantly higher blood levels of vitamins E and A than younger seniors. Another study in the Annals of the New York Academy of Sciences found that "adults aged 65 or older account for 60 percent of all cancers and 70 percent of all cancer deaths."

Another source of cancer-causing stress is meat from factory farms, which use hormones to induce animals to gain weight. Once ingested by us in the meat, those hormones can overstimulate our own hormones, leading to cancer. 12

PH Level Imbalance

The last method in which processed food harms us is perhaps the easiest to guard against. Everyone knows acids are corrosive and dangerous. No one would think of drinking a glass of acid, yet eating much of our modern diet causes the same effect inside our bodies.

Excess acidity decreases our body's ability to repair damaged cells, decreases our ability to absorb minerals and other nutrients, decreases energy production in the cells, decreases our ability to detoxify heavy metals, and weakens our immune system, making us more susceptible to fatigue and illness. Our bodies balance harmful acid by the intake of alkaline foods.

Measuring the acid/alkaline balance in our blood is accomplished by using a mathematical scale known as pH, or potential of Hydrogen. This scale starts at 0 for a totally acidic solution to 14 for a totally alkaline solution. A reading of 7.0 is neutral. Our ideal blood pH is a slightly alkaline reading of 7.4. Most Americans have an acidic reading below 7.0. Luckily, determining your body's pH level is easy through the use of pH strips or litmus paper.

Our ideal diet should consist of 80 percent alkaline forming foods and 20 percent acidic forming. Achieving that ratio is so difficult because most common foods are acid forming, including meat, poultry, seafood, dairy products, breads, grains, nuts, oils, and—especially ~ sweets. 13 This brings us back to our old enemy, sugar. Beyond being acid forming and, therefore, harmful to our body in general, sugar is cancer's actual fuel. The glucose in sugar is utilized by every form of cancer cell at a rate 10 to 50 times higher than that of healthy cells. 14 Cancer cells need massive amounts of glucose to reproduce and spread. Remember that the next time you guzzle a giant bottle of soda or chomp on a big hunk of candy.

Strangely enough, it turns out the stress Hans Selye so carefully studied has primarily become concentrated in one place, our dining table. Dr. C. Everett Koop, former U.S. Surgeon General, states that 75 percent of deaths in America are caused by avoidable nutritional factor diseases. Selye was way ahead of his time in realizing the devastating effect stress in all its forms has on our bodies. He left us with an insightful explanation of modern death:

To Die of Old Age

What makes me so certain that the natural human life-span is far in excess of the actual one is this:

Among all my autopsies (and I have performed well over one thousand), I have never seen a person who died of old age. In fact, I do not think anyone has ever died of old age yet. To permit this would be the ideal accomplishment of medical research ... To die of old age would mean that all the organs of the body had worn out proportionately, merely by having been used too long. This is never the case. We invariably die because one vital part has worn out too early in proportion to the rest of the body. Life, the biologic chain that holds our parts together, is only as strong as its weakest vital link." 15

Chapter 3—"WATER, WATER EVERYWHERE... BUT"

Without a doubt, the most glaring example of malignatech is modern man's attitude toward the freshwater which sustains the life of every person on Earth. Our ancient reverence for this beautiful, mysterious, most vital substance has degenerated into arrogant abuse.

We in America have become so alienated from Nature that we've forgotten the most basic fact of survival. We can live without oil and gas for the rest of our lives. We can live without food for weeks. WE CAN'T LIVE FOR A WEEK WITHOUT WATER!! After three days it's pretty much all over. Keep this fact in mind as you read this sad, disgraceful chapter.

From space our magnificent blue planet looks like the universe's ultimate tropical paradise. It very likely was, until technology arrived. Even though three quarters of Earth's surface is covered by water, man can access for drinking less than .08 of 1 percent of our world's total water supply. 1 To make the situation even worse, floods wash two-thirds of that available freshwater into the sea. 2

Humanity's water problem is two fold: a growing population competing for a diminishing supply. Mankind's demand for this resource is increasing at twice the rate of population growth. 3 This recipe for disaster kills over 10 million people a year through water-related diseases such as dysentery and cholera. Infants pay a heavy price, since polluted water is the leading cause of their death worldwide. 4 The United Nations' Population Fund predicts that by 2025 two-thirds of the world's population will experience a shortage of drinking water.

Compounding this problem are several malignatech blunders of epic proportion. The more technologically advanced a country, the bigger the blunders. Guess where that puts America?!

Waste

Americans tend to view freshwater as an endless waterfall cascading into our home and business. Our concept of the ultimate house is showers and sinks everywhere, a Jacuzzi in the bathtub, a hot tub on the deck, a gigantic pool in the back yard, fed by a towering waterfall. And don't forget the automatic sprinklers built into the lawn, garden hoses for soaking the plants. It's essentially free, so why not use all we want, right?

That is the simple reason we have yet to recognize any water crisis in this country. Clean water is still much too cheap. We pay an average of a quarter of a cent per gallon of drinking water. That cost will definitely have to increase as we are forced to rebuild our decaying water delivery systems. The Environmental Protection Agency has estimated that by 2025 America will need to have spent $277 billion just to *maintain* our drinking water infrastructure. 5 Add in the fact that we have the fastest growing population of any industrialized country, and things start to look serious. 6

Why is America the only country in the world to use freshwater to water lawns? Half the freshwater used in Florida is sprayed on lawns. 7 This wasteful practice will soon be a relic of the past.

We have over 700,000 miles of water pipes in this country. Water mains break approximately 237,000 times a year, because some of the pipe is over one hundred years old!

Worldwide, as much as 30—40 percent of water being transported is lost to canal or pipe leakage or illegal tapping. 8

Approximately 70 percent of all groundwater is used for agriculture, but an estimated 60 percent of that is lost to leakage or evaporation from inefficient irrigation systems. Thus, 42 percent — almost half!—of all groundwater used is wasted for its intended purpose. 9

Drip irrigation, in which water is delivered directly to plants' roots by underground perforated tubing, shows great promise for reducing evaporation losses from spray irrigation. Crop yields can be increased significantly, while using one third less water. Drip irrigation has been around since the 1970's, yet is being used on less than *one percent* of the planet's irrigated land! With all of its advantages, you would expect more farmers to be using it. Once again, logic and technology are strangers. 10

Mismanagement

As the guardian of Mother Nature's priceless natural resources, mankind is a failure. Malaxiom #2 sums up the problem:

> Malaxiom #2
>
> Technology was primarily developed to allow man to destroy Nature for profit.
>
> The more technologically advanced we become, the more we abuse Mother Nature.

For the past few years flooding has been constantly in the news. Hurricane Katrina almost washed away New Orleans. What are supposed to be "once a century" floods are now happening every other year. One obvious cause of urban flooding is the steady growth of our cities. As more land gets paved over, rain can't sink into the ground, so it gets washed into the sea, *after* it has flooded cities and caused billions in property damage and loss of life. Another widespread cause of flooding in America is the 79,000 dams we have erected over the years. These unnatural structures not only trap pollutants but also impede our rivers' and streams' natural flow. Approximately 3,500 of those dams are unsafe. Bringing all of America's dams up to full safety standards will cost $30 billion. 11 Is any government agency you know of working on this problem?

This country's most astounding instance of water mismanagement is our compulsion to move to the desert, an area not known for having plentiful water supplies. Some of our fastest growing cities, such as Las Vegas, Phoenix, and San Antonio, are in our driest climates. Las Vegas' regional water office has estimated that its metro area will run out of water as soon as 2020. That's based on current water usage. With the city's population projected to triple by 2050,2020 is probably a very optimistic date. Owens Lake, a former water source for Los Angeles, has totally dried up. 12

China and India, Earth's two most populous countries, both share an extremely serious problem, desertification. As the name implies, this is the scary process whereby cropland turns into barren desert. Improper irrigation, over-cultivation, and over-grazing are the causes here. Once fertile topsoil is blown or washed away, the process is difficult to reverse—unless you live in Las Vegas! China is especially at risk of desertification. Already 40 percent of Chinese land is desert, with about 2,500 square kilometers being added to this area annually! 13

Salinization is a major culprit here. This is the process in which natural salts in water evaporate as lakes are drained down. The resulting salt residue makes water unfit for fish or drinking. Russia's Aral Sea is a very sad example of salinization. Overdrawing of its waters in the past few decades has reduced the surface area by 50 percent. This increased salt concentration has destroyed a once-booming fishing industry. As the dried salt turns to powder, winds carry it to other land. Salt accumulating in soil ruins its fertility. Plants can no longer grow in it. With no plant roots to hold it in place, topsoil erodes and desertification begins. 14

Mismanagement of groundwater from underground aquifers is a worldwide problem. Since this water is usually cleaner than water found in lakes and is 30 times as abundant, the temptation is to just suck it up with abandon. In America we depend on groundwater for 50 percent of our drinking water. Not surprisingly, aquifers in Texas and California's San Joaquin Valley are being rapidly depleted.

Europe and Russia get 80 percent of their drinking water from aquifers. With groundwater depletion outpacing replenishment by an estimated 4 percent annually, the world is in trouble. Compounding the problem is the fact that in low rainfall areas the aquifer recharge rate runs to centuries! Mexico City, one of the world's most populous cities, could run out of water in the next decade. The entire country is in major water trouble. According to Conaqua, Mexico's National Water Commission, there has been a 50 percent drop in that country's water supply over the past 50 years! 15 Finally, coastal aquifers which are drawn too low can be inundated with salt water and ruined. 16

In 1977 the United Nations held the first major world conference on water. Now, over 30 years later, the worldwide water crisis continues to worsen. Why the lack of progress in such a vital area? Why do over half the developing world's water supply projects fail? 17

One shocking fact that highlights our collective stupidity is our destruction of natural wetlands. Since 1990 it has been estimated that half the world's wetlands have been destroyed. That figure is incredible, considering how vital these areas are to our water supply:

- They regulate river volumes
- They store and filter water
- They provide natural habitat for growing alluvial grasslands, which feed cattle, and rice that feeds people.
- Their stored water allows evaporation that maintains local humidity and rainfall. 18

Why can't all the world's concerned scientists, conservationists, and ordinary citizens make any progress in this ultimate survival challenge? Laziness, greed, and a strong dose of denial all play big roles here, but why are governments and large corporations doing virtually *nothing* to head off this impending disaster?

A 1996 report on drinking water by the U.S. EPA estimated that America's water system will need $140 billion to meet standards set by new laws such as the Clean Water Act and the Safe Drinking Water Act. YET THE FEDERAL GOVERNMENT HAS BEEN REDUCING ITS CONTRIBUTIONS TO LOCAL WATER SYSTEMS FOR THE PAST 30 YEARS!! Malignatech hard at work in Washington! Also, public funds are being used to subsidize corporate use of water, further worsening the crisis. For example, the city of Austin, Texas, charges companies water rates which are less than two-thirds of homeowner rates. 19

Pollution

As if waste and mismanagement weren't serious enough threats to Earth's freshwater, pollution provides the real sledgehammer blow.

Major polluters, governments, and large corporations conveniently forget that all of our planet's waters are interconnected. Pollutants dumped into a river will find their way to the sea, which covers the globe. It is estimated that two million tons of waste are dumped into the world's waters *daily. 20* How much longer can these waters absorb such an onslaught? Add in the number of chemicals involved and things get frightening. Of the more than 100,000 chemicals in everyday use in America, only a few *dozen* are monitored for safety! 21 Believe it or not, 1,600 chemicals have been detected in drinking water. 22

Even a seemingly normal combination of some of these chemicals can have deadly results. Our Mississippi River is flooded with hundreds of millions of tons of fertilizer from the feedlots and fields of the river's watershed (all of the Midwest and sections east and west of it). Remember the heavy metals in fertilizer I described in the last chapter? Those so-called nutrients are then funneled into the Gulf of Mexico, causing a huge algae bloom. This depletes almost all the water's oxygen content, resulting in widespread killing of marine life, especially shellfish and shrimp. Are we getting this lethal problem under control? Just the opposite! The 2007 summer dead zone was projected to be 8,543 *square miles*! 23 Our high tech society has deluded itself into thinking we dominate Nature to the point where we can abuse her at will. When that abuse takes the form of pollution, we are actually poisoning ourselves.

China's much-touted industrialization offers an excellent example of the toxic effects of accelerated growth. In addition to the previously mentioned growing desertification problem, China's environment is already one giant disaster. Since 2000 almost two-thirds of the country's 668 cities have had water shortages. Reservoirs, rivers, and lakes are all drying up. Around urban areas approximately 90 percent of these water sources have water that is unfit to drink. Of China's 700 major rivers, over half are polluted. 24 Throw in massive urban air pollution, and you have the makings of the largest environmental disaster in history. Where do 1.3 billion people turn when they run out of drinking water?!

American cities aren't much better off. The EPA has listed reservoirs serving such major cities as Boston, New York, Atlanta, and Dallas as becoming unusable in as little as a decade due to pollution.

It's important to remember that pollution doesn't just deny people drinking water. Third world residents often have no choice but to drink polluted water, since it is all they have. According to Worldwatch Magazine, waterborne germs and pollution kill 25 million people annually!

Global Weather Patterns

Global warming, global flooding, global storming. What's going on with our planet's weather? Some say the crazy extremes we've seen lately are being caused by our rapidly shifting north and south poles; some say ozone pollution is at fault; some say the culprit is sunspots. As serious as the cause is, the result of weather extremes is even more frightening.

Flooded areas get hit with a triple whammy. First, property is destroyed and lives are lost; then drinking water supplies in lakes and rivers are polluted by overflowing sewage and industrial waste from farmlands and paved surfaces; then the final insult comes when this massive downpour of drinking water washes out into the ocean instead of staying on land where it's desperately needed.

As some countries experience record floods, others receive less water, so severe drought sets in. Southeast Asia, ordinarily a lush region, is experiencing its worst drought in 40 years. 2007 marked the second year of the region's drought. The Asian Development Bank estimates the cost of damages at $20 billion. 26

Vietnam in 2007 received only 10 percent of its normal rainfall. Quang Tri province is suffering its worst drought in 90 years, according to officials. This lack of water is affecting 350,000 acres of rice paddies and costing hundreds of millions in crop damage. 27

Thailand, usually the world's largest rice exporter, didn't have enough water to plant a second crop in 2007.

In Malaysia's capital of Kuala Lumpur 600,000 residents have been affected by water rationing. 28

Even in the Philippines, one of the region's wettest countries, there is a water crisis. That country's Department of Environment and Natural Resources reports that of the 99 watershed areas in the Philippines 90 percent were so degraded they were "hydrologically critical." Causes of this disaster are destruction of forests, overgrazing, mining, agricultural expansion, and industrialization.

Privatization

As corporations have become larger and greedier, they have coined a new term to justify their attempt to take over the world. This so-called globalization has spawned another sinister term, privatization. When used in connection with drinking water, privatization means the transfer of ownership of water systems to private companies. Politicians like the concept because it allows them to dodge the issue of raising taxes to improve aging water systems.

There are three common forms of water privatization:

1) Complete sell-off of the water treatment and delivery system to a private company. This model didn't work well in Britain in the 1980's.

2) Leasing of water service to a corporation. This model led to major corruption in France, as companies attempted to bribe local officials to secure leases.

3) A government contracts with a corporation to manage water services for an administrative fee. The corporation here gets only a fee, no profits.

The inherent conflict-of-interest in privatization is that corporations focus on profits, not water quality or conservation. Also, privatization can lead to inequitable distribution of water to those better able to afford its higher price under this form of water ownership. 30

Privatization's most notorious failure happened in 1999 in the Bolivian city of Cochabamba. A London company, International Waters Ltd., bought the city's water systems. Incredibly, that contract actually gave the company control of all the city's groundwater, including residents' private wells!! International Waters attempted to raise water rates, ostensibly to pay for a proposed dam to provide drinking water, at a much higher price than alternative sources. Riots, martial law, and six dead demonstrators ensued. After water company officials fled the area in fear for their lives, the Bolivian government revoked International Waters' contract and passed a new national water law guaranteeing citizens more protection against greedy corporations. 31

With the global water industry now a $400 billion industry, the main objective of international water companies is to get as big a share of the pie as possible. Probably the most serious fault of privatization is that the poor often can't afford to pay for the more expensive water involved. Higher rates in South Africa in 1998 forced poor residents to get their water from polluted lakes and streams. Two years later a record cholera epidemic broke out. It lasted two years and infected 250,000 people, killing almost 300.

Large international water companies don't hesitate to use politics to further their cause. They often work closely with the World Bank to lobby governments to privatize public waterworks. By targeting poorer countries, the World Bank uses the leverage of loans to force governments into privatization. By 2002 the water companies had contracts with at least *56* countries.

In America these companies have moved aggressively to lobby Congress for corporate tax breaks. They're now attempting to persuade Congress to pass laws forcing municipal governments to consider privatization of their water utilities in exchange for federal loans and grants.

So far the United States has not warmed to the idea of entrusting our most precious resource to international water companies, for good reason. Higher rates, and thus profits, are the main goal of these businesses, as evidenced by their readiness to cut off water to anyone who cannot pay for this vital resource. Large water companies also promise to deliver better service at a lower price. The opposite often happens. Once the contract is secured, the companies quickly try to renegotiate higher rates and lower service standards. Public accountability is also weakened under privatization by the large companies' obsession with secrecy over contract terms and ongoing problems with water quality. 31

Will the world succumb to water privatization's siren song? As long as corrupt governments and irresponsible citizens continue to delude themselves into thinking ruthless corporations can be trusted with mankind's most vital resource, this disturbing trend may gain momentum. Its track record thus far is filled with evidence of higher costs to customers, environmental abuses, corruption, and inequitable distribution of water. A major wild card in this game is the globalization trend. Recent trade agreements, especially NAFTA, basically give rights to a country's natural resources to the highest bidder, regardless of whether or not that bidder is a foreign corporation. 32

In many ways the worldwide water crisis neatly sums up malignatech's insidious threat to civilization. Water represents the essence of Nature, its beauty, power, and vitality. By abusing and flagrantly wasting this priceless resource, modern man is demonstrating how detached he has become from Nature and her most basic dictates of survival. Since our bodies are composed of over half water, our abuse and waste are essentially abuse and waste of ourselves, Malignatech has provided the illusion of such power over Nature that we assume we can at last ignore her at will. Drain the ancient aquifers; technology will surely find more water. Pollute the rivers and oceans; the current will take the toxins far away. Waste billions of dollars searching for water on Mars; it makes for a great distraction so the masses don't notice our crumbling water infrastructure. Let greedy water bottling companies drain groundwater supplies; water is just another commodity to be bought and sold, right?

What does it say of our humanity when we allow technology to treat our Earth's most precious resource with such utter contempt?

Chapter 4—FRANKENFOODS HAVE ARRIVED!

Imagine a man in a business suit knocks on your door. He is there to inform you of a new kind of food his company is selling. He carries no samples or pictures of this new food. Instead, he explains that it isn't really natural and hadn't been tested on humans before being sold. Your eyes widen when he adds that the federal government hasn't bothered with any regulations or labeling requirements. He then casually admits people and the few animals willing to eat this new food have been made sick and even died from it.

You would probably be getting very turned off by this scary new food and tell the man you have no intention of eating such a toxic concoction. He replies you already have been, because this new food has been in stores since 1994 and about two-thirds of all processed foods contain it. 1 He then really floors you by stating that his company plans to take over the entire world's food supply with this toxic new food so everyone in America would be forced to eat it or starve.

What would you do after hearing this insane speech? Make a citizen's arrest or just beat him to a pulp? I have some bad news for you. This man's story is all true. The toxic, unnatural food he is promoting is called genetically modified food, GM for short. The GM story is a shameful disgrace to science and a frightening commentary on corporate greed and governmental stupidity.

How could some new kind of food cause such controversy? Today it seems we're reading more about genes than jeans. Corporate America has become obsessed with playing God by fooling around with genes in order to create new, *patentable* products that can make a fast buck. A gene is a chemical code that controls a particular attribute of a body, such as height or skin color. Genetic engineering involves transferring single genes from one organism to another to introduce that gene's characteristic to the new organism. 2 For example, potato growers wanted to create a potato which would have more resistance to cold, so they inserted in the plant a gene from a cold water fish! Sounds unnatural, doesn't it. That's because GM foods are just that, by definition. The U.S. Government's Genetic Modification Regulations define GM foods as "the altering of the genetic material in that organism in a way that does not occur naturally by mating or natural recombination or both."

GM foods' fatal flaw is that genes work in groups. They can't randomly be implanted in other organisms *without* disrupting the organism's natural genetic order and function. THUS, CHANGES IN HOST ORGANISMS CANNOT BE SAFELY PREDICTED!! This is especially true when anything toxic, such as pesticides, is added to plants. In 1989 Americans consumed the supplement L-tryptophan, which is derived from GM bacteria. As a result, 37 people died and 1,500 were permanently disabled! 3

In their mad rush to create patentable new products, GM food companies are showing zero regard for the present and future safety of our food supply. These so-called scientists assume that inserting genes in a plant will produce a single protein. Not true. Multiple proteins can be created, with unpredictable results. A plant's genetic structure may become unstable, creating rogue toxins. These adverse effects may not appear for HUNDREDS OF GENERATIONS after a crop has been genetically modified. 4 Rogue proteins can also be created here. Similar prions have been linked to mad cow disease and its human counterpart, Creutzfeldt-Jakob disease. Prions survive cooking and can incubate undetected in humans for up to 30 years!

It is extremely important to remember that these lab drones DO NOT know what they are doing! Even the most brilliant scientists don't understand living systems (such as the human body) completely enough to alter plants' genetic makeup without creating potentially harmful mutations. These mutations could infect *all* plants on Earth by the spread of GM seeds and pollen by winds, birds, and insects. The effects of gene pollution are *permanent.* GM food companies are being allowed to jeopardize the *entire* world's food supply with virtually no restrictions by the U.S. Government.

But what about all those colorful ads we see in magazines showing a smiling peasant with a large basket of glossy veggies? What about that new yellow GM rice with vitamin A spliced in to prevent blindness in the third world? Let's examine the GM food companies' claims for their products and see how they hold up.

Interestingly, GM foods are *not* tested on humans before being sold in stores. Animal trials with these frankenfoods have been shocking. Damaged immune systems, stomach and liver lesions, and sudden death are common. 6 Pre-cancerous cell growth has been found in test animals. Some animals are smarter than others. Hogs, cattle, wild deer, and raccoons in the Midwest refused to eat GM foods when offered them. 7

This wasn't the case in India, though, where cattle grazing on Monsanto's GM cotton plants were killed. Hundreds of Indian agricultural workers have reported allergic reactions to that same cotton. Some workers have even been hospitalized.

Here is a key point to remember about frankenfoods. The companies creating them know they are highly experimental, unnatural, and toxic, yet these companies are deliberately letting frankenfoods contaminate other crops, thereby *permanently* endangering our *entire* food supply:

- Unapproved GM long grain rice was discovered in August, 2006 to have contaminated the U.S. rice crop.
- In March, 2005 an unapproved GM corn variety escaped from Syngenta's field trials and contaminated U.S. corn.
- To save having to spray cotton plants with insecticide to protect them from pest infestation, Monsanto engineers isolated and inserted the gene of a natural insecticide into cotton plants. Now this cotton contains a toxic protein. Animals eat these plants; people eat the animals; guess what happens next! Remember, the FDA not only doesn't require human safety tests on GM foods before they are sold to the public, but the FDA also doesn't require any post-market surveillance of these foods! How many consumers will have to get sick or die before someone blows the whistle on this insanity?! 8
- The natural crossbreeding of GM foods that is occurring when GM pollen is blown onto another plant could create an accidental monster. For example, an herbicide-resistant crop could breed with a weed, creating a weed that is resistant to chemicals which used to kill it. 9
- In a December 29, 2005 report the USDA Office of Inspector General severely criticized the Agricultural Department for its terrible oversight of GM field trials, especially for high-risk drug-producing crops. 10

Product Safety

In 2005 Monsanto, the major producer of GM foods, paid a

$1.5 million fine to the U.S. Justice Department for bribing 140 Indonesian officials to allow Monsanto's GM cotton to be approved *without* an environmental impact study. Monsanto has also been found to repeatedly rig test data on GM foods to hide their dangers.

The World Health Organization and several other groups have expressed concern that if the antibiotic resistant marker genes used in GM foods got transferred to bacteria, the combination could create super-diseases which are immune to antibiotics! 11

Soon after GM soy was introduced to the United Kingdom, soy allergies soared by 50%. It turns out GM soy contains more of the most common allergen, trypsin inhibitor, than natural soy. 12

Recombinant bovine growth hormone (rBGH), a Monsanto product, is a genetically engineered growth hormone which can increase milk output in cows by as much as 10 to 30 percent. In other words, farmers get something for practically nothing. In this case, however, the ultimate price consumers pay could be their lives. In 1999 the United Nations Food Safety Agency banned the use of BGH after testing revealed that the GM hormone also increased levels of a growth factor which could pose a major risk of cancer. Did Monsanto respond by ceasing production of BGH? No. Health Canada also banned its use in 1999, but somehow BGH is still permitted in U.S. milk. 13

Who's Against GM Foods

It turns out the only people enthusiastic about frankenfoods are the ruthless companies concocting these freaks of Nature. Anyone who knows anything about them (with the exception of our brain-dead federal government) seems definitely opposed to GM foods:

- Four California counties have banned the planting of ALL GM crops, even approved varieties.
- More than 4500 jurisdictions in Europe and on other continents restrict the planting of GM crops. 14
- The Union of Concerned Scientists has expressed reservations about the dangers of GM foods.
- Lancet, the well-respected medical journal, states that GM foods should never have been allowed into the food chain. Britain's and Germany's Medical Associations have both issued similar statements.
- If GM foods are as safe as their makers claim, why has no major insurance company been willing to insure GM foods? 15 These people are the risk experts. What do they know about the dangers of frankenfoods?!
- Could this liability problem explain why GM food companies are so vocal about NOT labeling their toxic creations? Without a GM label, it's much more difficult to track down causes of sickness and disease in consumers. Yet surveys are finding that 85—90 percent of consumers want GM foods to be clearly labeled. 16 Right now the only GM food to be labeled is fresh fruits and vegetables. Their four digit PLU number has an 8 in front of it, making it a five digit code, such as 86543. Organic produce has a 9 as the first digit. This will be easy to remember. Just remind yourself (and everyone you know!) not to get caught behind the 8 ball. 17
- People with food allergies cannot trust any GM food to be safe for them. How does someone allergic to nuts know whether or not a GM food has had a nut protein added to it? Remember,

GM foods are not labeled.

- People sensitive to toxins in their food cannot safely eat GM foods. With some of these foods having pesticide bred into the actual plant used to make the food, consumers will be eating toxins that can't be washed off before cooking.

Frankenfoods are a horror movie come to life. These toxic, unnatural, untested, unlabeled, and unregulated products are a prime example of malignatech. It's scary enough to have greedy, ruthless corporations create these food substitutes, but to then see those companies use every deceptive trick in the book to ram this garbage down our throats is very terrifying. We all need to start *screaming* at our elected representatives and these corporations to stop this insanity! Remember that GM ingredients are in the majority of all processed foods in America. 18 Once you consume a frankenfood, you could have created a ticking time bomb within your body. The entire American public is being used as guinea pigs. This means you and your children. Any mysterious diseases any of us contract can't be traced back to the mad scientists who poisoned us. They've covered their tracks too well, thanks to major assistance from our corrupt, inept federal government. We could be staring at the biggest disaster in human history!

Chapter 5—THE CLONING NIGHTMARE

As insanely complex as the concept of GM foods is, cloning comes at us with far more complicated genetics, plus the added passions of religion and politics! If we survive having our food supply mangled and possibly destroyed (you'll love seaweed burgers and bamboo fries), malignatech has created what may become its biggest monster yet with cloning. Once again scientists feel their compelling obligation to play God by manipulating human and animal genes. Our big problem here is the same old one: all the scientists on Earth do not *fully* understand the ancient and incredibly complex mysteries of genes. Couple this ignorance with the overwhelming lure of power to create life and you have the makings of mankind's ultimate social nightmare.

If you doubt the intensity of that lure, look up cloning on your computer's search engine. There you will be informed there are 31 million websites dedicated to cloning! That works out to one site for every ten people in America. In short, cloning is a very hot topic.

There are three main types of cloning: DNA cloning, therapeutic cloning, and reproductive cloning:

- DNA cloning involves the transfer of a DNA fragment from one organism to a foreign host cell. 1 This technique of cloning a gene has been around since the 1970's. The process is too complicated to explain here, but the result is that billions of genes can be produced fairly quickly in controlled lab conditions.
- Therapeutic cloning is the highly controversial

creation of stem cells. This process entails replacing the genetic material in a human egg with a new nucleus. After a chemical stimulation ("calling Dr. Frankenstein"), the altered egg sometimes divides. After five days of dividing, the altered egg is mature enough to have stem cells extracted from it. This egg, which some people consider a tiny human being, is destroyed in the process. Researchers think this cloning procedure is worth that price because harvested stem cells can be used to create almost any type of specialized cell in the human body. These scientists hope stem cells can someday soon treat serious diseases like cancer and heart disease.

- Reproductive cloning is used to create an entire animal. This type is destined to become even more controversial than stem cells. Here the nucleus of an egg is replaced with the nucleus from an adult donor cell. This new egg must be artificially stimulated to grow into an embryo, which grows to a size suitable for transferring into the uterus of a female host. Dolly, a Finn Dorset sheep, was the first mammal to be successfully cloned. That was in 1996.2

Animal (reproductive) cloning has several problems:

1) Viability of our food supply—entire herds of cloned animals would be genetically identical, leaving them all vulnerable to being wiped out by a disease. Genetic diversity would be lacking.

2) Corporate secrecy—as with GM foods, big producers don't want any labeling requirements imposed on cloned products, with good reason. A recent Pew Initiative poll revealed that the majority of Americans are uneasy about eating cloned animals.

3) Why bother? Animal cloning is just more selective breeding to get an identical—supposedly better—product to sell. It's an expensive and unreliable technique. Cloned animals have higher rates of illness and sudden death. 3

If animal cloning can be done, why don't we start doing human cloning? Mainly because human cloning opens up *a gigantic* Pandora's box of ethical, medical, and financial issues. This subject is one of those extremely rare areas of technology where malignatech's dangers stand out in blinking neon letters one hundred feet high.

Amazingly, our U.S. Congress is considering legislation which would ban human cloning because of all its dangers. Fifteen states now have laws pertaining to human cloning. Thirteen of these states ban reproductive cloning. Six went even further by banning cloning for research purposes. 4

People are in favor of human cloning for some seemingly-valid reasons, such as curing cases of infertility and providing a means for gays and lesbians to reproduce. But, the obvious main reason is to try for immortality by creating a younger version of oneself. 5

All of the incredibly complex problems here are enough to drive even a trial lawyer crazy. Here are some of the major ones:

- Despite the fact that the human genome was sequenced in 2000, with all 25,000 of our genes identified, the challenge of truly understanding all of that complexity is mind boggling. 6
- Hidden defects can lie dormant and not appear until later in life. Some of those defects can cause sudden death.

- Would a human clone be considered a twin of the person or an offspring? Lineage problems abound here. Does the clone inherit the estate?
- Will cloned humans have the freedom to pursue their own destiny or be slaves to their creator?
- Will a clone be a real person with a soul?
- Remember the television series "Dark Angel," with Jessica Alba? She played a futuristic commando-ette who could leap long distances because she had been genetically engineered with some cat genes. The temptation for governments or corporations to combine human and animal genes will be very powerful. Is it possible that human babies could be born from other animals? 7
- Cloning definitely brings back memories of Hitler and his obsession with a "master race." Genetic engineering is still abhorrent to many people.
- Women would be exploited by human cloning. To induce women to give a quantity of eggs, scientists have to inject them with super-ovulating drugs. This procedure can lead to ovarian cancer and other health risks. Poor women would very likely be exploited here. 8

Considering the growing, toxic damage man has already inflicted upon our fragile planet, can we honestly expect man to play God responsibly? The Catholic Church obviously does not. It views any research which can harm human embryos or fetuses as immoral, since the Church considers these to be individuals. It also considers human cloning techniques contrary to the moral law since they are in opposition to the dignity of human procreation. 9

When science tries to play God, it should be a gigantic red flag to *all* of us, not just the Catholic Church! Of all the threats to our most important institution, traditional marriage, human cloning is probably the most serious. What better way to dehumanized us than to start creating what are essentially human robots! Why bother with the effort and expense of getting married when you can order offspring from a catalog? Think very carefully about how this trend will warp humanity. Class and race warfare could become the norm, with human rights getting lost in the colossal genetic reshuffling.

Keep in mind malaxiom #3:

> "Malignatech's goal is to dehumanize us to the point of self-destruction."

Chapter 6—MEET YOUR NEW BRAIN

In one of the most dramatic older Star Trek episodes, Dr. McCoy was faced with the challenge of giving Mr. Spock a brain transplant to save his life. Even though the story was set several centuries in the future, medical science still hadn't figured out how to do such a near-impossible operation. When some sort of supernatural force zapped the good doctor, he suddenly knew how to do it. "It's so simple a child could do it!" he exclaims in amazement. He then proceeds to save Spock's life.

That great episode was a telling commentary on our brain's future. Unfortunately for us, many Americans have already had a symbolic brain transplant; namely, letting computers do most of their thinking for them. They graduate from pocket calculators doing math to the internet providing theme papers and test answers. Then it's on to taking orders from a computer at work, their car's GPS system, and their television news at night.

Computer Vs. Brain

Some scientists estimate that artificial intelligence (AI) will be equal to the human brain in only a dozen years. This competition promises to be the battle of the century. Let's compare the contenders:

Computer—it's everywhere, in just about every size and shape. Small ones can give you the time and date on a digital watch.

Supercomputers can crunch a trillion calculations a second! Yes, that's a one followed by sixteen zeros.

Like the human body, a computer is composed of numerous components that each perform a specialized function. Through its stored memory this machine can perform logical operations which resemble human thinking. Aside from a power outage, the computer's key weakness is the data it's been given. If that info is incorrect, all of its calculations with be incorrect, also.

Human Brain—Comparing our brain to a computer is somewhat misleading. Our brain created computers to input, store, sort, and provide data to us. Our creator designed our brain to be a total survival device, one that can figure a way out of a hurricane at sea, then let us indulge in a lengthy discussion of Impressionist painting's portrayal of the natural world. In other words, our brains can do just about anything, with the list getting longer all the time.

The Ultimate Computer

How can three pounds of squishy flesh do so many complicated tasks? Its two key strengths are compressed power and specialization.

Its compressed power comes from a design even supercomputers can only dream of. Our brain's bio-electric operation involves a combination of electrical impulses and chemical transmitters. For a computer to duplicate this incredibly complex design would require it to have a separate brain controlling manufacture and release of these chemicals. It's unlikely computers will ever get that smart.

At birth our small brains contain an astounding one quadrillion (1,000,000,000,000,000) nerve connections (synapses)! This staggering amount provides a newborn baby with plenty of input power to be able to adapt to its environment. Consider all the powers a new brain has to acquire: the basic five senses of touch, taste, smell, sight, and hearing, plus reasoning, muscle control, and language. Just to make our survival even more of a challenge, the brain also has to control another set of secondary senses: balance, motion, sensing internal and body surface pain, temperature, and pressure. Those synapses become stronger or weaker depending upon the amount of stimulation they receive, which is especially critical during the first three years of life.

Our brain's bio-electric structure is influenced by signals flowing in from all parts of the body and by hormones. Sex hormones play a major role in shaping the brain, with males leaning toward spatial abilities while females favor social skills such as language.

By age ten a person's brain has lost about half of its synapses, leaving 500 trillion to last for most of his or her life. The lost ones die off from lack of use. That decade of adapting to our world sets the stage for our brain's ultimate role, survival. Our small eyes can process information at computer-like speed. 1 When it comes time to do computations, the brain can perform massive parallel (simultaneous) computations with ease. You actually have a supercomputer sitting on your shoulders!

A quick look at specialization within our brain reveals a design complex enough to stagger the imagination. There are four main areas:

Cerebrum—this largest part takes up most of the skull's brain cavity. The cerebrum is split into two hemispheres that are closely linked. Different sections, or lobes, perform different functions. Our frontal lobe controls speech, emotion, and thought. Behind that lies the parietal lobe which receives and interprets sensations like pain, touch, and temperature. Behind the parietal lobe in the back center of the cerebrum lies the occipital lobe whose job is detecting and interpreting visual images. Yes, we actually "see" in the darkness deep inside our brain. Finally, on either side of the cerebrum sit the temporal lobes which store memory and control hearing.

Cerebellum—the brain's second largest part sits underneath the back of the cerebrum. This section coordinates muscles and controls balance and posture.

Diencephalon—lies atop the brain stem beneath the cerebrum's middle. Two key structures are housed here, the thalamus and the hypothalamus. Our thalamus serves as a router of incoming nerve impulses, directing them throughout the brain so that it knows what's happening in the outside world. Our hypothalamus handles conditions inside the body by regulating thirst, hunger, body temperature and by controlling the pituitary gland's release of hormones. All this activity occurs with no conscious effort on our part.

Brain Stem—this small structure rests atop the spinal cord where it regulates such vital functions as breathing, heart rate, sleeping, and waking.

Our entire brain structure is surrounded by a dense cerebrospinal fluid which also protects the spinal cord from blows and other shocks. Encasing all of these parts is our sturdy skull, made even stronger by its powerful curves.

Most Americans don't realize that their brain is composed of 60 percent fat, much of which is the omega-3 fat, DHA. Our modern diets have substituted omega-6 fats (from corn, soy, and sunflower) for omega-3 fats. When our brain tries to use these inferior fats to repair nerve cell membranes, faulty membranes are created. Neurotransmitters can't properly engage with these faulty membranes. As a result, our body cannot tell itself to release natural anti-depressants like dopamine and serotonin. The final result is increased depression and violent behavior.

How can this be? Bad diet leads to violence? Could this help explain the epidemic of school shootings America has experienced recently? Statistics say "yes."

Those have been provided by a psychiatrist and physician with the National Institutes of Health. Dr. Joseph Hibbeln has spent several years studying the effects of omega-3 fatty acids, specifically the effects of deficiency of these vital fatty acids. What he discovered is amazing! Dr. Hibbeln studied statistics for 38 countries since the 1960's, comparing the increased consumption of omega-6 fatty acids to any rise in depression and murder rates during the same period. In *every* single case there was a direct correlation! An increase in omega-6 consumption *always* results in an increase in depression and murders.

Modern antidepressant drugs such as Paxil, Prozac, and

Zoloft were supposed to cure depression. These selective serotonin reuptake inhibitors (SSRIs) increase serotonin levels in the brain, but come with a long list of serious side effects like headaches, nightmares, nausea, hair loss, constant fatigue, and weight gain. A recent study also revealed that these pills produced such a small improvement in most patients that it was not clinically significant. To top it all off, SSRIs can't correct the omega-3 fatty acid, trace mineral, and vitamin deficiencies associated with depression. Nor do these drugs treat the unresolved stress. 2 What they essentially do is cap the volcano until it erupts.

As America's diet gets worse, expect to see more depression and suicidal behavior, such as self-mutilation. In a recent Psychological Medicine survey of 633 high school students, 46 percent—almost half!—admitted cutting, biting, or burning their skin. They explained that this mutilating gives them a sense of control over their lives and that the pain distracts them from stress in their lives!! When teens become so warped they lose respect for their own bodies, you *know* malignatech is winning the battle for their minds.

With computers getting faster, smaller, and cheaper all the time, why don't we rely more on them for all our many personal needs? Why memorize information when computers have almost unlimited memory capacity? Let's do things the easy way, right?! Wrong. That lazy way may be the direction we are headed, but it's a fatal mistake. By not using *all* of our abilities to full extent, we are severely weakening our ability to survive this increasingly crazy world. Remember that our brain is the ultimate survival computer. It's a souped-up Porsche which most people just use to visit the grocery store. It yearns to have you take it out to a desert highway and turn that baby loose!

But you say, there's that blinding computational speed and vast memory. We created them; why not use them? Our body has a simple answer to this inevitable question. Those two seeming weaknesses of our brain, speed and memory, are viewed by our body as irrelevant. Human survival has never required super computational or memory powers. It has required *using* our brain. Remember those 500 trillion synapses we lose as babies because they weren't used? The more we use those synapses, the stronger they become. Areas of our brain which get massive use can get larger to accommodate the increased work load. Someday gyms may have to install "brain workout" rooms to keep us smart.

Unlike today's computers, with all their constantly changing designs and software, our brain has one simple purpose: keep our body running as efficiently as possible. To do this the human brain has developed an array of powers that leave *all* computers in the dust! Subconsciously, our five senses receive information from the environment and process it to keep us safe and healthy. Consciously, our senses also process information from the outside, then think of how to use this data. Somewhere between our subconscious and conscious worlds lies a growing area of powers we're just beginning to harness: instinct, precognition, remote viewing, past lives, and other abilities which used to be considered either demonic or way out in left field.

An interesting development in this area of expanding brain power is an old concept which is getting new scrutiny. "Embodied cognition," building on decades of human movement research, essentially postulates that our body is an extension of our mind. Studies have confirmed that solving problems and learning are enhanced by certain gestures or body movements. This concept is an important reminder that we used to be dynamic creatures who were constantly performing some sort of manual labor.

"Embodied cognition" has been around for a surprisingly long time without being taken very seriously. Early in the last century Italian educator Maria Montessori set up her now-famous schools designed to teach children in a dynamic environment which emphasized manipulating physical objects. Blocks and cubes were used to teach math and sentences were acted out by children.

It's easy to see how such a fun, creative learning system never caught on with America's "prison" concept of public schools. Our kids have always been herded through schools like cattle with pens and notebooks.

Considering how complicated our body is and the complex movements it is capable of, "embodied cognition's" elegant simplicity makes very good sense. Why did it take so many centuries to discover that our brain is not just a fleshy motor or computer? Rene Descartes' theory of mind-body separation had been the generally accepted one until recently. By the 1980's a group of scholars had the audacity to point out how our bio-electric brains did not process information in the same way electrical computers do. A series of studies have since verified this assumption:

> In 2000 Cornell psychology professor Michael Spivey and his student assistant made a fascinating discovery about how people's eyes helped solve a tricky spatial relations problem. These researchers noticed subjects given this brainteaser exhibited a distinctive and unconscious eye movement pattern just before they discovered the answer. It was as though their eyes were being used to walk through the possible solutions.
>
> Then an August, 2007 study by two University of Illinois psychologists carried Spivey's study further. Alejandro Lleras and Laura Thomas repeated the brainteaser test, this time inducing the eye movements Spivey had discovered. These researchers found that this step significantly improved the success rate for people attempting to solve this problem. Most of them had no idea how their eye movements were involved here. Lleras even learned that his subjects

thought the eye movements were a distraction intended to prevent them from solving the problem.

More evidence our body is an extension of our brain can be found in a study involving math problem solving. Susan Goldin-Meadow, a psychology professor at the University of Chicago, gave children arithmetic problems which normally would have been too difficult for them. She discovered the kids were more likely to get the correct answer if they were told to gesture while thinking.

There are skeptics who dismiss "embodied cognition" as merely another scientific fad. It seems too simple to have any deep meaning about brain operation. The tests mentioned previously, however, hint strongly that our mind-body connection is far more complicated than modern psychology realizes. Considering the epidemic of obesity and inactivity sweeping America, "embodied cognition" should be an area of urgent study. 3

High Cost of Low Parental Involvement

Today's complicated, stressful living demands optimal brain development of our children. Yet judging by how far their test scores are below those of their peers in many other industrialized nations, something is wrong here. SAT scores of college-bound students peaked in 1964! 4 Once again, one of the big Three R's comes into play. Parental involvement in their children's mental development is faltering due to the ongoing disintegration of our two-parent American family. A recent nationwide study showed one child in five (20 percent) under age eighteen has an emotional, learning, behavioral, or developmental problem related to not being in a two-parent family.

Forty-two percent of American families with children have at least one of the following strikes against them:

> Strike One—lack of education. The mother has her first baby before finishing high school.
>
> Strike Two—lack of commitment. The two parents are unmarried when the first child is born.
>
> Strike Three—lack of maturity. The mother is under age twenty when she has her first child.

Sadly, 9 percent of new families have all three strikes against them. One of the most shocking recent discoveries is that environmental stress can actually activate genes linked to depression and other mental illnesses. The price society pays in lost productivity and high health care and prison costs is enormous. 5

At the beginning of 2008 U.S. jails and prisons held a record 2,319,000 adults. In 2007 corrections costs for our 50 states—not counting federal expenditures—was over $49 billion, the most of any country on Earth. How many low-cost houses, food, and medicine could $49 billion buy!? The rate of prison cost increase in the last 20 years is 6 times greater than that for higher education. 6

Fortunately, scientists have recently learned how to use our five senses to create positive experiences which can program our brains to be smarter, healthier, and less violent. There is, however, a narrow window of ideal age to induce these positive physical changes in the brain: between the ages of 4 and 10. Not stimulating young brains is an expense America can no longer afford. Seventy-five percent of all American males in prison have low IQs and poor school records. 6

Brain Study Futility

Modern man's jealousy of the incredible genius that created our brain is becoming more and more desperate. We possess monster supercomputers capable of performing a trillion computations a second, so why can't we duplicate the operation of our tiny, three-pound wonder? How can such a small device continue to reveal powers we never realized it possessed? This isn't fair! Malignatech dictates that man is supposed to be able to dominate *all* of Nature, even his own body. Will we ever have the humility and intelligence to admit we will probably *never* completely understand how our remarkable brain functions?

Judging by modem science's ham-handed approach to studying the human brain, the answer is no. What I call the demon factor (electronic devices that emit radiation) enters the picture. Neuroscientists (brain specialists) who used to be confined to standing outside our skulls and taking educated guesses, are now armed with computerized imaging devices which can study the brain in amazing detail. Here are the big two:

Functional MRI (fMRI)—magnetic resonance imaging's use of magnetic rays doesn't subject a person to any radiation. The "functional" term refers to MRI's ability to picture brain activity over a period of time, from seconds to minutes. By measuring such brain functions as electrical discharges and changes in blood flow, an MRI can pinpoint either the location of a brain activity or the duration of it. Interestingly, neuroscientists studying an activity can pinpoint the location to within millimeters and the starting time to within one tenth of a second, yet no technique can do both.

Once again, our brain is just too complicated. Brain activity can't be tracked by measuring one neuron at a time because the process involves thousands of neurons firing together.

CAT Scan—this combination of X-rays and computer technology was invented in 1972 by Godfrey Hounsfield and Allen Cormack. In a CAT scan the X-ray beam moves around the body part being examined, creating multiple views or "slices." These highly detailed images are put together by computer into a three-dimensional picture that is light years ahead of the old X-ray. CAT scan has proven to be an invaluable tool for diagnosing disorders *in emergency situations.* The new 64-slice CAT scanner can take thousands of images in a matter of seconds, allowing doctors to see a body part or the whole body in amazing detail.

That's the end of any good news about CAT scans. As with most modern electronic devices, CAT scanners produce toxic radiation. Unlike those other devices (television sets, microwave ovens, cell phones, etc.), this radiation is serious:

- CAT scan's ionizing radiation is a proven carcinogen.
- The annual number of these exams has soared from 3 million in 1980 to an estimated 62 million in 2008. During that 28 year span cancer has become an epidemic. Is this a coincidence?
- At its highest power a CAT scan can produce more radiation than 1,000 chest X-rays!!
- It's important to remember that radiation exposure

is cumulative in the human body and much more damaging in children than adults.

- Since CAT scanners are very expensive machines, hospitals and clinics want to make them profitable, so these businesses charge thousands of dollars for whole-body scans, then promote them heavily as a wise preventive tool.
- There is no proof that these whole-body CAT scans are medically useful to healthy people. Even the FDA agrees, by stating "...the FDA knows of no data demonstrating that whole-body CAT screening is effective in detecting any particular disease early enough for the disease to be managed, treated, or cured..."
- Even when CAT scans do discover abnormalities, doctors often recommend further tests which subject patients to even more radiation and possibly dangerous operations of dubious value.

CAT scans are classic malignatech. Let a big, expensive, very complicated machine dazzle the masses with an exciting new test which does more harm than good. All of the hopeful hoopla shoves rationality aside, as America mistakenly assumes (for the umpteenth time!) that any fancy new technology can do no harm. Once again, the old way of health promotion was more effective. Back when doctors actually spent more than a few minutes with a patient, they would take a good history, give a thorough examination, then encourage the patient to make any appropriate lifestyle changes. Machines never have and never will care whether you stay healthy or alive.

Although the computer wins the hype part of this contest, our brain is and very likely always will be the ultimate computer. Remember, though, that your brain's power depends on its being used often. While use wears out electronic computers, it only makes your brain stronger.

Chapter 7—A NATION OF ADDICTS, THE LEGAL KIND

Nowhere in America is malignatech more prevalent *and* more deadly than in our conventional medical system. This dark, modern disease has corrupted our once-beneficial system into a monster whose goal now is to get and keep Americans sick. Money and power rule modern medicine.

This may seem an extreme claim, but I will prove it by the end of this chapter. First, however, a confession is in order. I approach this chapter with more than a little anger. Conventional medicine has put me through an expensive, painful, 32 year long ordeal that almost killed me. The maddening part is that most of it was unnecessary. Back in the fall of 1976 I had the first of 8 attacks of atrial fibrillation, a heart rhythm problem in which the smaller, upper chambers of the heart lose their normal rhythm and start quivering. This slows the blood flow to the larger, lower chambers, the ventricles, and affects one's entire body blood circulation. I lost a major portion of my strength and endurance, plus ran the risk of having all that blood sloshing around inside the atria pooling long enough to form a blood clot which could have caused either a heart attack or a stroke. It required a couple of doctors to finally discover why I felt so exhausted all the time. Atrial fib is common among the elderly but not in healthy, athletic 32 year olds. Over the next 32 years I had seven more major attacks, coming every several years. A series of cardiologists prescribed numerous toxic heart medications and blood thinners. According to *all* of them there was no known cause. My atrial fib was considered "paroxysmal," meaning it just happened periodically for no reason.

I always suspected there was a simple reason and cure for

this terrifying condition. It ruined my promising career as an advertising copywriter and kept me in constant fear of getting overheated or overexerted. About 10 years ago I decided to study this condition until I found that simple cure. After my 2001 attack I vowed never to go through another. That and all of my previous attacks had forced me to undergo cardioversion, which is exactly like on the hospital shows. Get put under, get zapped several times with the shocking paddles to the chest, and hope all the brutality coaxes your heart back into a normal rhythm. I was lucky to wake up alive after each operation. Something could have easily gone wrong, and I wouldn't have wakened up to see the usual circles burned onto my chest.

My research led me to alternative medicine and its emphasis on honest, natural cures. One day a few years ago I finally hit the jackpot. My favorite newsletter. Alternatives, arrived with an article about America's epidemic of magnesium deficiency. I learned that our constant use of synthetic fertilizers and overuse of our soils have resulted in widespread magnesium-deficient soil. Then I read that the latest U.S. government study shows a shocking 68 percent of Americans don't consume the recommended daily intake (RDI) of magnesium. 1 According to the National Institutes of Health, magnesium is vital in more than 300 biochemical reactions in the body. This magical mineral has also been found to support a healthy immune system.

I also learned magnesium is an important element in maintaining normal heart rhythm. Bells went off, whistles screamed, and I went crazy! It appeared I had found my cure. I upped my daily intake of magnesium from a few hundred milligrams to an even thousand. My heart has been in great shape ever since.

What does my experience say about conventional medicine? First, it says that an alternative health newsletter knows more about curing heart ailments than most cardiologists in

America. Secondly, it says that all those "board-certified" quacks don't really understand how the human heart functions. We are entrusting our lives and a big chunk of our fortunes to so-called professionals WHO DONT KNOW WHAT THEY'RE DOING!! They don't care about *preventing* health problems. If we don't get sick and stay sick, they can't drive their Jaguars, sail their yachts, and enjoy those vacation homes.

I've been burned—literally and figuratively—by our insane conventional medical system. Read this chapter carefully so, hopefully, you will avoid becoming another victim.

The Era of Stress

Back in the late 1940's an historic shift in American medicine occurred. Brilliant breakthroughs in vaccines all but eradicated the last of the serious diseases terrorizing this country. Smallpox and polio, then measles were finally cured. Antibiotics largely replaced natural cures in the fight against infection. It appeared that modern science was about to conquer all the major diseases.

Now, a half century later, that ideal world is a faded dream. Americans constitute only 5 percent of the world's population, but we consume 50 percent of all prescription medicine. 2 Despite a constant parade of new drugs and treatments, 52 infectious diseases are attacking us now in America. Among the new ones are AIDS, Lyme disease, silicosis, and acute pesticide poisoning. 3 The pace is quickening. According to the World Health Organization, at least 30 new diseases have appeared worldwide in the last 20 years. Some of the more notorious: hantavirus, salmonella, E coli, Ebola virus, Lassa fever, Legionnaire's disease, "mad cow" disease, and toxic shock syndrome.

What went wrong?

During the last fifty years our excessive faith in antibiotics and vaccines led to a reduction in public health spending on prevention and detection programs. This lapse allowed infectious diseases to establish themselves all over the western world. 4 Plus, the overuse of antibiotics has enabled diseases to mutate around these drugs. Your winter flu shot formula is a guess at which strain of flu will hit this season.

My referring to malignatech as a disease may seem contrived, but it reveals all the classic signs of a serious disease. It begins insidiously with multiple appeals to our baser instincts: laziness, greed, gluttony, power, etc. Then, once it has infiltrated our mind and body, malignatech begins destroying both. Laziness leads to weak bones, creaky joints, and general weakness. Greed gets us collecting more and more money to buy more status symbols. Gluttony leads to obesity and ailments related to poor diet. Power leads to wars, coups, and complex technology very few people can understand, such as nuclear power. The destruction feeds upon itself.

In addition to polluting our topsoil with toxic industrial waste, we've ruined much of our tapwater. The EPA estimates 45,000,000 Americans drink tapwater contaminated with chemicals, lead, germs, parasites, bacteria, etc. Even more disgusting is the Center for Disease Control's (CDC) statement that nearly 85 percent of Americans have parasites in our bodies! When Nature isn't attacking our bodies, our own toxins are. The average American is exposed to 50,000 chemical agents in his or her lifetime. Only a fraction of these have been studied for their effects on humans. 5

This massive pollution assault on the U.S. population strikes hardest at the weakest among us; namely, our children. According to the National Center for Health Statistics 31 percent of American children have a chronic health problem! That's almost one out of three kids—unbelievable! In the past 20 years childhood autism has increased at a rate ranging from 200 to 7,000 percent. During that time childhood diabetes has tripled, while the rates for asthma, learning disabilities, and Attention Deficit Hyperactive Disorder (ADHD) have doubled. The biggest growth of asthma cases has been in children under age 5.

What is going on here!? Why should supposedly healthy American kids be getting hammered by so many diseases? Does the term "canary in the mine shaft" ring a bell? Our children, who trust adults to keep them healthy, are being betrayed by a conventional medical system which puts mercury, the *most* toxic heavy metal, into their vaccines and dental fillings, then throws other dangerous drugs at any minor aliment. Antibiotics wipe out good and bad bacteria. Our bodies need the good to fight illness.

Unfortunately, the timing for our next generation is terrible. At the same time they are being abused by conventional medicine, many new drug-resistant strains of bacteria are appearing all over the world. A mutant strain of E.coli is being blamed for over 9,400 cases of food poisoning, two-thirds of them children. Great Britain was hit with a drug-resistant strain of salmonella, sickening 4,000 people. That episode has been directly linked to the use of antibiotics in livestock. 6

In another tragic example of how malignatech feeds upon itself, there is now a lethal, antibiotic-resistant form of bacteria that is attacking patients IN hospitals!! You read that correctly; people are dying from an infection they catch AFTER being admitted to the hospital! This bizarre story began in 1988 when a formerly easily-controlled bacterium named enterococci developed a resistance to vancomycin, the antibiotic that had been able to easily control it. Strains of vancomycin-resistant enterococci (VRE) started showing up in patients being treated in a New York City hospital. In the next three years the number of hospitals reporting cases of

VRE jumped from one to 38. Of the 100 infections reported 98 were found to have occurred *after* the patients had been admitted to the hospital. VRE infections spread to states surrounding New York, then to Europe, then all over the world. In all of the infected countries a similar infection pattern developed:

1) The longer the hospital stay the greater the chance of developing a VRE infection.
2) Patients previously treated with vancomycin or other antibiotics have an *increased* risk of becoming infected.
3) Most patients developing a VRE infection are 41 to 60 years old.
4) When the infection gets into a patient's bloodstream, death almost always results. Drugs cannot stop VRE infections.

Interestingly, natural remedies have been shown to cure VRE infection. 7 Have viruses and bacteria finally outsmarted modern medicine? We'd better hope not, or "pandemic" may become a household word soon.

In a classic case of too little, too late, new laws are being passed mandating public reporting by hospitals of their hospital-acquired infection rates. Seven states have passed such laws, with 20 other states considering doing so. Finally, prospective patients will be able to compare hospitals on this vital statistic. Considering that each infection averted saves a hospital an estimated $25,000 in unreimbursed costs, one would have to ask why these institutions have been so slow to address such a deadly problem. 8

What is wrong with modern drugs? More importantly, what is wrong with us to expect a pill to cure every ailment we have. Overweight? Find a pill that will take those pounds off. You might remember that popular one, FenPhen. It helped with weight loss but also caused heart valve defects and was recalled in 1997. Drugs by their very design are unnatural substances. Why do we expect our bodies to take them in and not suffer side effects, possibly serious ones? Our culture has become lazy, impatient, and amazingly short-sighted with regard to pharmaceutical drugs.

You don't have to be a mathematician to do the drug numbers. A lot of people take more than one drug, all of which interact with the food they eat. Now add in even a couple of vitamin supplements, and the possible interactions of all these substances becomes a number so big and complicated no one knows what's happening. A University of Toronto study found that approximately 2,216,000 U.S. hospital patients experience a serious adverse drug reaction annually. Of those reactions 106,000 are fatal, which makes adverse drug reactions the fourth leading cause of death in America. Remember, those 106,000 people are patients dying *in* hospitals! What does that say about the effectiveness of pharmaceutical drugs?

Despite over a dozen recalls of drugs in the last few years, the public still doesn't see the truth of what a colossal scam these medications have become. Malignatech has invaded every aspect of this industry.

Need For Drugs?

With all the new diseases assaulting America, there isn't much news about cures for any of them. This is all part of the new direction of modern medicine, which is treating problems and managing symptoms. When you've created a $500,000,000,000 industry (yes, that's *billion),* why cut your profits by trying to cure diseases?! This is where the Hippocratic Oath of "first, do no harm" has gone out the window. Rather than concentrate on curing deadly diseases, modern medicine has actually focused on *creating* phony diseases which just happen to be treatable with their expensive, toxic, often ineffective drugs!! This betrayal of the public trust is a national disgrace that history will deal with very harshly. Baldness, low self-esteem, overeating, and even aging are now classified as diseases! 9 Perhaps the two most blatant examples of this insanity are cholesterol and osteopenia.

I'll never forget my first exposure to the cholesterol-lowering scam, many years ago. There, just inside the door of my local supermarket, was a neat table attended by two ladies in white uniforms. At first I assumed they were screening donors for a blood drive, but they informed me they were screening cholesterol levels, looking for people at risk of having high numbers. This was all news to me. The reason I walked quickly past the table was my instinctive reaction to this numbers game. As a former EMT, I know the medical profession's love of "magic" numbers which only they control. Any numbers I can't do at home (blood pressure, pulse, etc.) are highly suspect to me. It turns out my suspicions about cholesterol numbers were correct. Far from being an evil substance, cholesterol is an essential building block for our bodies. Our livers manufacture plenty of it, which our bodies use to make healthy cell walls and hormones. Our brains are mostly cholesterol. Those "magic" numbers don't add up. Despite America's turn toward a low fat, high fiber diet in the last 30 years and gradually lowering bad cholesterol numbers, heart disease has become an epidemic. Why the disparity? Research has proven that those numbers are essentially meaningless. In 1992 the British Medical Journal reviewed 30 years of major studies and concluded, "lowering cholesterol does not reduce mortality and is unlikely to prevent coronary heart disease." The report also stated that a significant number of people develop heart disease *after* beginning cholesterol-lowering treatment!! In a 1995 report the New England Journal of Medicine stated "most patients with myocardial infarction (degeneration or inflammation of the heart) have normal cholesterol levels."

If this mysterious fatty substance isn't causing heart disease, what is? The answer was discovered 40 years ago but, as is the custom with the medical establishment, was not only completely ignored but also ridiculed. In 1968 a Harvard University pathologist named Dr. Kilmer McCully was researching the problem of clogged arteries. He discovered that people dying of this problem had high levels of homocysteine in their blood. This amino acid has a corrosive effect on the artery linings. Excessive amounts of homocysteine in the bloodstream create pits and ridges inside arteries. Those rough spots catch fat and cholesterol flowing through and create deposits, which over time can build up enough to block blood flow. Dr. McCully's brilliant research has finally been validated by a New England Journal of Medicine study that found high homocysteine is a risk factor 14.4 times greater than high cholesterol. The medical establishment's blaming cholesterol for heart disease turns out to be a case of blaming the Band-Aid for causing the cut because it happened to be in the area. Why was the cause of much of our heart disease ignored for 40 years? Very simply, because the cure was a natural one, which couldn't be patented and sold in an expensive pill. Excess homocysteine is caused by a lack of certain vitamins and nutrients. 10 Our modern diets are at fault, again. This goes back to Dr. Koop's statement about avoidable nutritional factor diseases.

Now that the true cause of clogged arteries is public knowledge, you might expect cholesterol-lowering drug manufacturers to back off promoting their statin drugs. You would be wrong! They are not about to give up their cash cow. Taking a cue from that old adage "the best defense is a good offense," the drug companies are now recommending even lower LDL (bad) cholesterol numbers for most people. This would give them at least 6 or 7 million new customers.

The objective here is obviously to sell as many statin drugs as possible before the general public really wakes up to this scam. Statins' risk-reward ratio makes even less sense than those "magic" numbers. Why should a pill that only controls—doesn't cure—a phony disease cost an average $3 a pop?! Why do so many people get hooked on these drugs when they have so many serious side effects? Remember Baycol, a popular statin which was recalled in 2001 after it was found to cause kidney failure and death in some users? Why are the other statins still on the market when they can cause the same problems? In addition to fatigue and serious liver damage, statins hinder the body's usage of such vital heart-protective nutrients as coenzyme Q10 and niacin. Plus, these drugs have been shown to increase your risk of dying from cancer and pancreatitis by a scary 44 percent! 11 Natural cholesterol-lowering nutrients, which can't be patented by you-know-who, are ignored or actively suppressed by the FDA. 12 Sadly, the public isn't being told that these natural remedies are far cheaper, just as effective, and carry zero side effects. Watch the news about statins. As the scam gets more exposure, you will see these toxic drugs promoted more and more as all-purpose cures for everything from baldness to being short. Anything to keep those big bucks rolling in.

The reason I mention natural cholesterol-lowering nutrients at all is because there is a chance that in one segment of our population high cholesterol *might* be connected to heart disease. This segment consists of men aged 30 to 55 and derives from the famous Framingham Heart Study, begun in 1948, in which over 5,000 adults were tracked for decades. Since this age group of men doesn't have a lot of heart attacks, the need for anyone to lower his or her cholesterol really does not exist.

A similar new "disease" has been created with bone density. Is it a coincidence that this scam has come along just when all of the baby boomers are reaching retirement age and getting creaky in the joints? You decide. A recent meeting of bone experts set some "magic" numbers (sound familiar?) for bone density for everyone, regardless of age or body size. Now if your bone density score comes in between -1 and -2.5, you're automatically labeled as suffering from bone loss. By another odd coincidence, half of all post-menopausal women fit into this new category of having osteopenia! Of course the cure being touted is another dangerous drug these women are expected to take for the rest of their lives. Once again, there are natural cures which are cheaper and more effective. Studies have shown these supplements cut bone loss by 26 percent in only one year. 13

While epidemics of VRE and heart disease sweep the globe, our drug companies are hard at work inventing new diseases and expensive pills that "manage" the symptoms. Curing diseases is so old fashioned, and unprofitable. Malignatech's insanity knows no bounds.

FDA and Drug Testing

The Food and Drug Administration's story is the darkest chapter in this book and the most evil chapter in American history since slavery. You will see how malignatech's frightening power turned a regulatory agency into a terrorist group, funded by *your* tax dollars. All of the abuses I am about to describe have been well documented, so rest assured I'm not making any of this up.

This agency was founded a century ago to ensure the safety of our food and drugs. Things went routinely until that pivotal era of the 1950's when antibiotics came on the scene. Natural medicine was marginalized as drug companies rushed to patent their inventions and start raking in vast profits. As those profits rose, so did intimidation and harassment of natural healing providers.

In 1962 the FDA introduced new "safety and efficacy" regulations for the testing of drugs. The accompanying increased costs had the effect of reducing the number of new drugs coming onto the market. These new regulations also had obvious flaws. While the approval process became more drawn out and expensive, it did *not* become safer! More people started dying of dangerous drugs after 1962. 14

Prior to 1962 the average approval time for a new drug was 7 months. By 1967 that time had increased to 30 months. Before 1992 the time averaged just over 7 years! How many people have died while waiting for a life-saving drug to be approved?

Finally, in 1992 the Prescription Drug User Fee Act was enacted to speed up the drug approval process. Approval times dropped from an average of 27 months in 1993 to 14 months in 2001. From one extreme to another. Predictably, the number of drug recalls during that period tripled. 15 How about efficacy? Are all these expensive pharmaceutical drugs really curing us? No! More people are dying of degenerative diseases now than in the 1950's.

As drug approvals became more and more expensive, the FDA became more dependent on the drug companies' millions of dollars. The American Medical Association and various government agencies have joined the FDA in a brutal terrorist campaign against anyone who threatens drug company profits. S.W.A.T. teams are actually being used to terrorize innocent natural remedy providers. Our "War On Drugs" (which has been a spectacular failure) provides a convenient excuse to seize property on the mere suspicion of illegal drug ties. The accused then has to go to court and "prove" his or her innocence!! This insanity is obviously unconstitutional, but how many times have you seen or heard any mention of these governmental crimes in the news media?

One of the most blatant and violent of these raids occurred in 1992 against Dr. Jonathan Wright, an internationally-known nutritional medical expert. Armed FDA agents broke down his clinic's *unlocked* door looking for contaminated vitamins, which they didn't find. Workers were terrorized at gun-point while agents carted away books, files, computers, and mailing lists for 14 hours. No charges were filed against Dr. Wright, who has sued the FDA over this insane attack. Welcome to the world of malignatech!

Stop and think carefully about what these flagrant violations of our constitutional rights mean to you. Your tax dollars are going to support an agency which terrorizes natural health providers, yet *never* raids the offices of major drug companies, even after their toxic pills actually kill thousands of Americans. Through its outrageous crimes the FDA is openly admitting they are owned by the pharmaceutical companies, the very businesses this agency is supposed to be regulating.

As further proof of this last statement, consider the story of aspartame, the artificial sweetener. The FDA's Food Division had refused to approve aspartame's use back in the 1970's because it produced brain tumors and seizures in lab animals. In 1981 it was finally approved only because President Ronald Reagan, a friend of Searle, the manufacturer, took office and fired the FDA Commissioner who wouldn't approve aspartame. Dr. Arthur Hayes, the new commissioner, then went so far as to overrule an FDA Board of Inquiry that still didn't want this toxic sweetener approved. Soon after aspartame was approved for use in carbonated beverages, Dr. Hayes left the FDA for a position with—you guessed it!—G.D. Searle Co.

When aspartame is heated above 86 degrees F, as in cooking a product, it releases methanol, the same toxic fluid you add to you car's gas tank to prevent gas line freezing. In 1996 the FDA removed all restrictions from aspartame, allowing it to be used in all baked and heated foods. This methanol is broken down in the human body into formic acid and formaldehyde, a deadly neuro-toxin. Adverse reactions to aspartame include brain cancer, mental confusion, depression, insomnia, migraine headaches, nausea, vision loss, death, and about 80 other ailments.

If you think all this is shocking, hang on. It gets *much* worse! During Desert Storm in Kuwait our troops drank large amounts of aspartame-sweetened sodas that had been heated over 86 degrees in the desert heat. Remember the Gulf War Syndrome, when many troops returned home with symptoms of formaldehyde poisoning? 16 Our government never seemed to find a cause for that syndrome, did it? Now you know why. It gets even worse. Two-thirds of the children born to those troops had birth defects!

Once again America sends its troops off to an ill-advised war, then *deliberately* poisons them!!

In Vietnam the poison was Agent Orange, a carcinogenic defoliant sprayed on the jungles our troops walked through. In Desert Storm the poison was toxic soda that never should have been made. Your tax dollars are funding an agency of the federal government that is *deliberately* allowing you to eat poisoned food. Malignatech really is killing us.

Another danger of pharmaceutical drugs that their manufacturers and the FDA do not publicize is drugs' ability to deplete the body of vital nutrients. It is estimated that 30 percent of pharmaceutical drugs' side effects are the direct result of drug-induced nutrient deficiencies. That "temporary relief everyone craves can thus lead to permanent damage. For example, taking Tetracycline, a popular antibiotic, can deplete your supply of, interfere with activity of, or increase you need for *sixteen* nutrients. Some of these are folic acid, riboflavin, vitamin C, magnesium, zinc, and calcium. How about aluminum antacids, which are used to neutralize excess stomach acid? Aside from the aluminum part (which has been linked to Alzheimer's disease), they sound harmless enough. Not so. They can deplete your body of *nine* essential nutrients: folic acid, vitamins A and D, calcium, chromium, iron, magnesium, phosphorus, and zinc. 17 Prednisone, used to treat inflammatory ailments in various parts of the body, is another major nutrient depleter. It can take away *eleven* nutrients, including vitamins A, B6, C, D, and K, folic acid, and magnesium. When essential nutrients like these are depleted, the side effects could be serious. What's the best "cure" for a serious ailment? Another drug, of course. Do you see a vicious cycle here? Could this cycle be the reason why older people are on so many prescription drugs? Why sell someone one drug when you can hook them on four or five, right?

Why aren't serious side effects more prominently noted on drug packages? There's an old sales adage which goes "don't introduce any negatives into the sales pitch." You can find mention of some of these side effects if you want to squint through two or three pages of the mouseprint that always follows a magazine ad, but should users have to be detectives?! A General Accounting Office report found that over *half* of patented drugs approved by the FDA between 1978 and 1986 had "serious post-approval risks" not disclosed on their original packages. By "serious," the GAO meant risks such as heart, kidney, and liver failure, blindness, seizures, and birth defects. Once more, why aren't pharmaceutical drug company offices being raided by armed FDA agents?

This agency's drug approval process is still a mess. Most clinical drug trials don't run long enough or involve enough people to identify rare side effects. There is usually inadequate testing on various subgroups, such as women and children and people with serious ailments (of the heart, liver, etc.). Interactions are not addressed. A drug can have a toxic effect when combined with other drugs or supplements. 18

Increasingly, FDA approval of a new drug carries less credibility among industry groups. Many health plans now don't consider using a drug until it has been on the market for at least six months. 19 Even that delay is very minimal, considering that it took four years for Baycol to be recalled and five years for Vioxx. Public Citizen, an advocacy organization, recommends waiting seven years unless a drug is a rare breakthrough medication.

In their rush to sell more drugs to more groups of people, the industry began advertising directly to consumers in the 1980's. This blatantly commercial ploy encourages use of drugs by people who may not be appropriate candidates and also downplays serious side effects. The FDA relaxed ad guidelines in 1997 and will probably continue to do so. This advertising has been very profitable. Consumer groups advocate two or three year moratoriums on ads for new drugs to allow serious side effects to be discovered before tens of millions of Americans start taking a possibly toxic medication.

Showing Its True Colors

The real duty of any governmental agency is to protect the lives of American citizens, the people paying all those salaries. FEMA's recent failure to do that with Hurricane Katrina could be chalked up to normal federal red tape and incompetence. With the FDA, however, there is NO excuse for its many recent failures to protect American lives. Here are some of the most outrageous:

- There was almost a *two year* delay between the time Merck warned the FDA about serious heart risks it had discovered with Vioxx and the FDA's changing the warning label. In the end Merck recalled the drug, not the FDA.

- America's desperate hunger for quick cures has led to an explosion of new drugs in the last few decades. Although the FDA tries not to approve drugs with names that sound similar to already-patented drugs, mistakes happen. If a similar-sounding name gets misread by a pharmacist because an arrogant doctor couldn't be bothered to print legibly on his prescription pad, a serious or fatal reaction could occur. Such a case happened in 2004, when the maker of Advicor sued the maker of Altocor for having a too-similar sounding name. To top it off, both were cholesterol-lowering drugs. Altocor was renamed Altoprev, and the maker agreed to pay Advicor's manufacturer a $6,000,000 settlement, for a mistake the FDA could have easily caught. 20

- According to the Journal of the American Medical Association the third leading cause of death in America is conventional medicine *itself*!! An unbelievable total of 225,000 deaths a year are caused by medical mistakes! Nonerror, adverse effects of medicines are responsible for 106,000 of those deaths. Why does the FDA allow this to happen? Why aren't drugs more rigidly screened for toxic side effects? Why aren't consumers better informed about potentially fatal side effects of drugs? The other numbers may surprise you: 12,000 deaths from unnecessary surgery, 80,000 deaths from hospital infections (remember VRE?), 7,000 deaths from medical errors in hospitals, 20,000 deaths from other hospital errors. Do you get the feeling hospitals are very unhealthy places?!

- As prices for patented drugs soar, Americans are increasingly turning to generic versions. We now spend $30,000,000,000 a year on generics. As some widely-used drugs have their patents expire soon, that amount is expected to rise. There is now a backlog of 800 applications for generic drugs at the FDA, twice the number of 2003. Lack of staff and delaying tactics from brand name manufacturers (such as lawsuits over patent violations) are slowing our access to cheaper drugs. 21
- Our FDA's arrogance at protecting drug manufacturers from competition by natural remedies is astounding. Despite several *thousand years* of evidence to the contrary, this pathetic agency has declared that only drugs are capable of preventing, treating, or curing a disease. If a substance does any of those things, it's considered a drug and must be tested and approved by guess who! The fact that most drugs are derived from natural substances doesn't enter this brain-dead mindset.

Cherries are the FDA's latest target as a drug they should be regulating. The controversy began with a $141,210 grant from the U.S. Department of Agriculture to the cherry industry. This grant was to allow the industry to investigate cherry's health benefits and promote sales. It turns out there were some amazing benefits in these small fruit. Studies funded by the grant determined that cherries rate high on the ORAC antioxidant scale, are fat and sodium-free, are high in folic acid and vitamins B6, C, and E. A 2004 study by Johns Hopkins Hospital discovered that cherry's red pigment, called anthocyanins, suppresses inflammation as well as popular anti-inflammatory drugs. A 2001 study by a leading researcher reported that cherries contain fairly high levels of melatonin, a natural sleep factor now also know to be a factor in immunity. Cherries, it appears, are a miracle fruit. Logic would dictate that the cherry industry should be permitted to promote their fruit's health benefits since those benefits have been documented in reputable studies. The FDA has its own definition of logic and insists only that agency has the right to know which studies can be revealed to the public. In cherry's case, the FDA does not accept any of the studies. Your tax dollars are going to pay for studies which this agency, also funded by your dollars, deliberately suppresses. Vital health information is being withheld from you by an agency which believes it can redefine truth!

Fortunately, the U.S. Supreme Court is on our side here. It has repeatedly slapped down the FDA's "highly paternalistic approach" to withholding information from the public. In a verdict on the FDA's attempt to suppress alternative medicine, the Court stated: "Bans against truthful, nonmisleading commercial speech usually rest solely on the offensive assumption that the public will respond "irrationally" to the truth. The First Amendment directs us to be especially skeptical of regulations that seek to keep people in the dark for what the government perceives to be their own good." 22

How would you like to participate in the testing of a risky new drug, *without* being asked whether or not you wanted to?! Incredibly enough, this is what your FDA is allowing Northfield Laboratories Inc. to do to 360 badly injured trauma patients in a recent 720-patient trial. An experimental blood substitute called Polyheme (plastic blood, perhaps?) is being injected into half the hemorrhaging trauma patients while the other half gets a saline solution in the ambulance, then donor blood when at the hospital. This insane violation of patients' right to know what drug they are being given is all the more criminal considering that Polyheme flunked a previous, smaller trial. That was an 81-patient study of aneurysm surgery (bulging of the artery wall). In that trial, 10 of the 81 patients given Polyheme suffered heart attacks, two of which were fatal! None of the patients on standard therapy had heart attacks. Another federal agency, the Office for Human Research Protections, has complained to the FDA, expressing "urgent ethical concerns" about the conduct of this trauma study. Once again arrogance leads a company to think it can invent a substitute for human blood, probably the most complicated and essential substance in the human body. 23

As corrupt and inefficient as the FDA is, there are some glimmers of hope that its power is diminishing. A recent federal appeals court ruling stated that dietary supplements can carry health claims as long as they attach a disclaimer that those claims have not received FDA approval. 24 With such approval getting more suspect all the time, that ruling is a big victory for natural remedies. If you watch weekend infomercials on television, you've probably noticed there are more of them lately, and their claims are getting bolder. Perhaps an era of truth in medicine is dawning!

Americans appear to be wising up, with more now turning to natural health remedies. A recent New England Journal of Medicine study discovered that about one third of American adults use unconventional medical treatments, such as chiropractic, relaxation techniques, and megavitamins.

LATE BREAKING NEWS: Several months after I finished writing this bizarre chapter, something even more bizarre happened. The FDA actually admitted it was *incompetent*!

In response to mounting criticism of his agency, FDA commissioner Dr. Eschenbach asked a special committee to evaluate the FDA's competence. The resulting 60-page report repeatedly stated "The FDA cannot fulfill its mission because...." it had inadequate staff, technology, and organization. Perhaps most shocking was the statement "The FDA does not have the capacity to ensure the safety of food for the nation." Remember my chapter on frankenfoods, genetically modified food? In all the years this toxic food substitute has been stealthily introduced into our food supply, the FDA now admits it "cannot even keep up with the advances in science." 25 Next time you buy food or a prescription drug remember you are on your own as far as safety goes.

This latest failure of a powerful, well-established federal agency makes an important point about how out of control technology has become. Today's explosive changes in all areas of this field are malignatech's perfect breeding ground.

CHAPTER 8—A NATION OF ADDICTS, THE ILLEGAL KIND

As the last chapter documented, legal drugs are becoming a toxic plague. While many Americans aren't aware how dangerous they are, *everyone* knows the deadliness of illegal drugs like steroids, cocaine, LSD, and all those designer drugs. Thanks to the magic of modern chemistry, this list is constantly getting longer.

Before we delve into that murky world, let's examine an extremely toxic substance that should be illegal but isn't.

The Curse of Mercury

It's the most lethal non-radioactive substance on Earth. Mercury is also unusual in being a liquid metal. Although this strange metal is now used primarily in the manufacture of industrial chemicals and in electronic applications, mercury has an ancient history of use in tonics, makeup, and medicines (laxatives, antiseptics, and even antidepressants!). Those of you who believe the conventional medical industry is trying to get us sick and keep us sick will see massive evidence here that mercury is the smoking gun.

Why would any health practitioner want to put such a deadly metal into the human body? One answer is because mercury is liquid and, therefore, easily compounded with other metals. When mercury amalgams (dental fillings) were introduced in America in 1833, there were two different types of dentists. "Craftsmen dentists" had no formal medical training and often worked as barbers or blacksmiths. "Medical dentists" were doctors who had training in both medicine and dentistry. This group was aware of mercury's toxicity and used gold to fill cavities in teeth. The craftsmen dentists liked mercury amalgams (made of 50 percent mercury, along with silver, copper, tin, and zinc) because they were cheaper and easier to work with.

These two groups formed opposing dental organizations, the American Society of Dental Surgeons (ASDS) and the American Dental Association (ADA). The ASDS banned mercury amalgams in 1843, considering their use as a form of malpractice. When the controversy finally led to the ASDS' collapse, the ADA emerged as the country's dominant dental organization, which it still is. 1

Although the ADA still insists dental amalgams are safe, it recently admitted that mercury vapors are released from dental amalgams in the course of chewing. A Swedish corrosion PhD, Jaro Pleva, analyzed a five-year old filling and discovered that almost half the mercury had leached out. So much for the "hard and stable" theory. Once ingested into the body, mercury can pass into the brain and wreak amazing havoc. There are over 125 symptoms of mercury poisoning, including depression, asthma, and vision problems. Our EPA considers a dental amalgam removed from your mouth to be toxic waste! There is a set disposal procedure which must be followed to protect dental personnel from mercury poisoning. Does this insanity remind you of the fertilizer scam, where toxic waste changes its name depending upon where it's put?! Several European countries (Sweden, Denmark, and Germany) are considering banning mercury amalgams. 2

Here's where the smoking gun starts blazing! Everyone is familiar with childhood vaccinations. From the 1930's until recently an organic compound called thimerosal has been used in vaccines to prevent bacterial contamination. Few people were aware that thimerosal is 46 percent mercury.

Why was mercury, in ANY amount, used in a medicine, especially one given to vulnerable infants?!!

Many people suspect the dramatic increase in autism cases since the late 1980's is directly linked to mercury in vaccines. Parents cite the finding that fully vaccinated children have been receiving in their first two years mercury levels which exceed safety limits set by the FDA. Autistic children tested for mercury have revealed excessive amounts of the toxic metal. What better time to get us sick than when we're infants. That pretty much guarantees we'll be sick *all* our lives! There is a special place in hell for these crooks.

All this may sound very paranoid. Why would our federal government, with all its supposed watchdog agencies, be an accomplice to a plot to poison our children? The simple answer is money and lucrative jobs with drug companies. The proof of this is a secret meeting held in Norcross, Georgia in June, 2000. Top government officials from the Centers for Disease Control and Prevention (CDC) and the FDA, high-level vaccine specialists from the World Health Organization (WHO), and representatives from every major vaccine manufacturer quietly met to discuss the use of thimerosal.

This meeting's primary concern was a new study which raised troubling questions about the safety of a number of childhood vaccines containing this disgusting heavy metal preservative. Tests had indicated thimerosal was responsible for dramatic increases in autism in children, as well as a number of neurological disorders such as attention-deficit disorder, hyperactivity, and speed delays. These officials showed their true colors by spending the next two days plotting how to cover up the shocking data.

The CDC (funded by your tax dollars!) led the charge to hide the facts:

- They paid the Institutes of Medicine to produce a new study to disprove that this obviously toxic preservative could cause brain disorders!
- They withheld damaging reports by claiming the information had been permanently lost.
- To prevent the discovery of the facts through the Freedom of Information Act, the CDC handed its huge database of vaccine records to a private company, making the information unavailable to the public or researchers! 4

Despite its very obvious toxicity, thimerosal is still allowed in numerous over-the-counter medications by the FDA. Would that agency allow even a tiny amount of lead to be added to any medication? Of course not. Look at the outcry over lead in Chinese-made toys. Why then, does the FDA allow mercury, which is *more* toxic than lead, to be put in medications?!!

There is only ONE reason your federal government would allow an extremely lethal substance such as mercury to be added to any medicine. That reason is perfectly clear: to make you sick, so that drug companies (which essentially own the FDA) can sell you even more medicine. This insane poisoning of Americans, especially our children, should result in life sentences at hard labor for all officials and scientists involved!!

Parents, wake up! Your children are being *deliberately* poisoned. If any state or federal agency demands that your kids be vaccinated for ANY reason (including flu shots), demand proof that 1) the vaccine contains NO mercury of any kind and 2) the vaccine actually works (which rules out useless flu vaccines). Do NOT assume any group—including the federal government—has the right to poison your children! A great idea would be to start class action lawsuits against vaccine manufacturers and any governmental agency backing them.

Consider the evidence against flu shots. According to the media, all manner of flu viruses are about to rain down on us: swine flu, asian flu, avian flu, you name it. Therefore, everyone must have an annual flu shot, right? Hogwash! Here's what you *aren't* being told about flu shots:

- Thimerosal has been almost completely removed from children's vaccines, but is *still* being used in flu vaccines. Why?!
- Studies have shown that flu shots are marginally effective.
- Flu shots cause some people to contract a paralyzing Guillain-Barre type syndrome of temporary—and sometimes permanent—paralysis!
- Tens of thousands of flu shot recipients end up with the flu, anyway.
- The pharmaceutical industry vehemently objects to Americans buying prescription drugs from Canada. Their stated reason is because these drugs may be counterfeits made in foreign countries which do not adhere to America's strict quality control standards. What the drug industry conveniently forgets to mention is the 2004 flu vaccine shortage caused by a California company having to destroy 48,000,000 doses of contaminated flu vaccine. It had been manufactured *not* in America, but in Liverpool, England. How do you spell hypocrisy?! 5
- Here's the clincher. In some hospitals more than half the doctors and nurses refuse to get flu shots! Why would professionals who are constantly

> exposed to flu germs do such a thing? Could it be they know the truth? 6

You don't have to be Sherlock Holmes to put all this damning evidence together: toxic industrial waste being put in our fertilizer; poisons being put in our food; the *most* lethal heavy metal on Earth being put in our medicines. It's very obvious lots of companies are out to poison the American people. Get us sick, keep us sick, sell us assorted phony "cures," make billions of dollars.

Wake up, America!! You and your children are being *deliberately* poisoned! Question the safety *of everything* you put into your body.

Designer Death

Everyone knows street drugs are dangerous. Nobody but a loser would take such illegal and dangerous stuff. Change the name to designer drugs, and malignatech suddenly kicks in. Since designer jeans are elegant and cool, then designer drugs must be, too, right? As the saying goes, "it's all in the marketing."

Not everyone knows that heroin was the first designer drug. It was synthesized around the turn of the century as a less addictive form of the popular pain reliever, morphine. Most modern designer drugs are analogs (variations) of heroin, amphetamines, or PCP. 7 The concept began as a way to circumvent narcotics laws by creating synthetic versions of illegal drugs which are different enough not to be illegal. Congress slammed the door on that scam in 1970 by passing the Controlled Substance Analog Act, which outlawed all designer drugs. Needless to say, the law didn't stop dealers from concocting a constant stream of new drugs. The dual lure of big profits and relative safety of not having a supplier who can turn you in to the feds both keep small drug labs sprouting up all over the country.

Just as heroin replaced morphine, new drugs known as fentanyls could be heroin's successor. They look and act like heroin but can be thousands of times stronger, yet their high lasts only 1 to 2 hours. With such strength, the safe and fatal doses are very close together.

Unlike designer jeans made by a fashion house, designer drugs have no verifiable quality control. Their purity and strength are unknown to the buyer, as are possible toxic interactions with other substances, such as alcohol. In giving the user a high, these drugs disrupt neurotransmitter production in the brain. This leads to an increased tolerance of the drugs by the brain but not by the rest of the body (heart, lungs, etc.). As a result, possible reactions to designer drugs can be severe: seized-up chest muscles, total paralysis, permanent brain damage, and death.

So why would a sane person pay good money for such a risky high? A major reason is the old standby, peer pressure. Designer drugs morph into club drugs when they're used at all-night dance parties known as raves, held at dance clubs, bars, and empty buildings. In such a seemingly safe social setting drug users are inclined to want to dance all night, since peak effects of drugs such as ecstasy can last for 8 hours. Add alcohol consumption, which causes dehydration, which can lead to overheating caused by the extended exertion of dancing, and you have a vicious downward spiral into sickness and possibly death. It's a seductive ritual of "feel good" drugs, pounding techno-music, flashing laser lights, and glow sticks. All those high tech trappings hide one scary fact: designer drugs are extremely dangerous. One use can cause permanent brain damage.

Muscles To Die For

Everyone wants a body like Arnold's, but very few people want to earn a great physique by years of strenuous exercise and careful diet. There must be an easy way to bulk up.

Modern science to the rescue with steroids! This class of drugs was developed in the 1930's to treat conditions arising when the male body produces abnormally low amounts of testosterone. Steroids, which are taken either orally or injected, used to be legally available only by prescription. Now they're being sold on many internet sites and even being put in over-the-counter medicines for asthma and inflammation. Children and babies are being given these medications, which can cause *immediate and permanent* damage. 9

Steroids are dangerous because they are powerful hormones which have dozens of dangerous—sometimes fatal—side effects. Anabolic (growth) steroids are synthetic versions of testosterone, which activate cells at the genetic level. Once these specific growth proteins are created, the travel throughout the body. This is when undesirable side effects can occur. Importantly, these side effects many not require months or even years to develop.

There are more than 70 side effects, some of which can be fatal. A partial list includes

liver and kidney tumors

jaundice

high blood pressure

trembling

extreme mood swings ('roid rage)

insomnia

heart disease

Importantly, these problems can lead steroid users to turn to other harmful drugs to alleviate the side effects. 10

Control acts passed in 1990 and 2004 have not been helped by revelations that pro athletes such as baseball stars have used steroids. Adolescents desperate to grow up quickly are thus tempted to follow this bad example. Kids as young as eighth graders are abusing steroids. According to a 2003 University of Michigan study, 4 percent of high school seniors have tried steroids at least once. Ironically, the side effects to teens can be devastating: stunted growth, severe acne, increased hostility, and long term damage to fertility and sexual functions. Steroid abuse is far more sinister than other teen vices such as smoking. Even if a user decides to forego muscle bulk by stopping steroid use, most of the side effects will have caused *irreversible* damage. 11

As a true example of malignatech, steroids also do serious damage to America's social fabric. The recent Mitchell report about steroid use in major league baseball has hammered that traditional sport's all-American image. How does all this tainted play affect the aspirations of kids dreaming of baseball stardom? Why bother working hard to develop playing skills when other guys are getting theirs from illegal drugs? How do aspiring female athletes feel when they see Olympic stars like Marion Jones finally admit to illegal drug use and turn in five gold medals? With more sports stars being discovered as cheats, why should kids have heroes anymore? Honesty, hard work, integrity.... what good are they in a world which only rewards results? As malignatech steadily destroys rationality, we can be prepared for more and more scandals to further weaken our already weakened sense of outrage. Do you see a downhill progression here?

CHAPTER 9—THE SLOW DEATH OF ART

What is the soul of America? Is it our religions, with all their many traditional values? Or is it our capitalist drive for money and all the glory it can buy? There is very likely only one feature of American culture that touches the lives of all of us, regardless of our religion or our wealth. That would be the amazing variety of art we produce.

Let's examine how technology has affected those forms which are most dependent on it: music, television, and movies.

First, though, an important bit of background information. I'm 65 years old, which means I actually grew up as a little kid not watching television, because they weren't being sold until I was about seven years old. Also, my timing for rock 'n roll was perfect. The year I turned 13, 1956, was the year Elvis came on the scene and launched rock into the stratosphere. This all gives me a historical perspective most Americans don't have.

Music

Speaking of Elvis, did you know he won three Grammy awards, all for his gospel songs, and not a single one for his immortal rock songs? Oddly enough, after his death the music industry then somehow came to its senses and voted him Entertainer of the Century! Jealousy, ignorance, stupidity? Why didn't the industry acknowledge his genius sooner? The important point to remember here is that industries and corporations are basically a collection of very fallible people.

Our music industry is probably the best example of that fact. The songs we listen to and sing reflect our concerns and emotions. In the post-World War II '50's America suddenly realized it was top dog in the world with the only nuclear weapons around. Then the Soaring Sixties brought the sexual revolution with the Pill, women's lib, and the drive for racial equality. By adding the "demon factor" of electricity to stodgy old guitars and keyboards, technology gave music a power it had always dreamed of. Now a small band or even a single guitarist could get up on a stage in front of thousands of people, plug into some monster speakers, turn on the juice, and totally blow the audience out of its seats! Music could finally kick ass!

Unfortunately, that power quickly became abused. Bands competed for the biggest speakers and loudest sounds. People left concerts with ears ringing. Society's fringe elements decided they wanted that power, so we got acid rock, punk rock, hate rock, etc. Popular music slowly transformed itself from something to soothe the savage beast, to something to arouse the savage beast.

Soon music found many new outlets. From radio, concerts, and movies, music wove its way into television, Walkmans and iPods, then into computers. Protest songs became popular. Everyone wanted to use music's vast new power to reach people. Music became a partisan tool, and in the process saw its beauty and optimism seriously eroded.

The most blatant recent example of music's abuse has been that pathetic ghetto ranting known as rap. A bunch of angry black men grabbed mics and began ranting about people they hated, which turned out to be everyone but themselves. They hated women and gays and cops and all whites and politicians and rich businessmen, etc., etc. Someone forgot to tell these no-talent losers that hate has never been an art form and, hopefully, never will. Luckily, a lot of the warped teenagers who used to enjoy this garbage are wising up, and rap is slowly heading where it belongs, into the trash can of history. The saddest fact about rap is how badly it trashes black music's magnificent heritage. From the old ballads to Count Bassie's big band sound to our fabulous modern women: Ella, Aretha, Diana, Whitney, Beyonce. From Nat King Cole (probably the smoothest voice of all time) to Smokey, Stevie, Michael, and Marvin. Why would a race with such an amazing depth of brilliant musical talent allow those angry ranters to trash their precious heritage? Is phony black pride more important than very real black heritage?

As disturbing as rap's hateful wordings are, its bizarre glorification of criminal behavior is even more warped. The "gangsta" rapper image these ranters try to present is pure malignatech. Toss rationality out the window and go for the sleaze. Does being a so-called celebrity mean you're allowed—or required—to act like a mindless criminal: killing people who give you a hard time, getting in shootouts at nightclubs, doing drugs, etc.?

Most disturbing of all is the general public's acceptance of this sick music substitute. Notice how few people in or outside the music business are willing to blow the whistle on this scam. Why? Why are Americans so easily lead astray by anything new, loud, and shocking? Why do we accept album covers with bloody body parts on them? Where is the outrage against this garbage?!

Rock's early unsavory variations, punk, hate, death, acid, were bad enough. Most of us were probably relieved to see all that trashy rock fade from popularity in the last couple of decades. Unfortunately, the hate and anger there seem to have caused a permanent infection in much of pop music. Rock's actual physical power to deafen us gradually transformed it into a weapon. Excess became the norm, as malignatech took over a beautiful art form to give us deafening concerts with wild lighting and special effects, lip syncing singers (Milli Vanilli won a Grammy for doing it!), mixing and editing of recordings, throbbing techno beats, etc. Basically, the glitz became the dominant theme, as one look at the outrageous group, KISS, will convince anyone.

All these problems with pop music may seem academic. With pop culture increasingly materialistic and violent, why shouldn't pop music reflect all that? It should—to a point. My contention is we have gone beyond the sane point and seriously weakened the essence of music, its incredible power to soothe and enlighten us. In short, to elevate us above our savage selves *not* celebrate the savage within. Remember a very, very critical fact about music in today's crazy world: these magical sounds are this planet's *only* universal means of communication. They transcend culture, language, age, wealth, and religion. Put a hand over your heart, feel that steady beat, and know that genuine music is the essence of your life.

Television

Ever notice an interesting new television trend? As Americans turn more toward computers and iPods for viewing, television is getting less attention from the critics. One very real reason for television's lack of critical attention may be that, with the new parental guidance ratings popping up at each show's beginning, everyone assumes the violence and sex problems have been solved. True, parents can now know a show's content ahead of time, but kids are watching more television without parental supervision. The average home now has more television sets than people, almost three sets. Plus, the ratings have provided a clever cover for even more sex and violence. Tipper Gore, where have you gone?!

Once again rationality is being bent to the breaking point by money-hungry networks. With more channels popping up all the time, competition for viewers is intensifying. Since sex and violence is a profitable combination, networks keep throwing more and more of it at us—all in the name of profits for their shareholders, of course. The resulting brutality is clearly warping American minds. Mafia murderers are "deeply troubled" men supposedly worthy of our sympathy. Serial murderers are brilliant minds who have just gone astray. Murderous golddiggers are merely status seeker women who are too pushy. The violent drug dealer is just a businessman protecting his sales territory. Rationalizations for brutal and illegal behavior are everywhere.

A bizarre new television genre is playing with dead bodies. People cut them up to painstakingly determine how they were killed. Other people revive them for a last shot at mortality. Still other people talk to the ghosts of the dead. Never has being dead been so cool.

Lost in all this glitz and big guns is the simple fact that murder and torture and betrayal are all dehumanizing. Young kids are having their minds warped. School shootings are getting more frequent and deadlier. Elaborate new rationalizations are constantly being concocted by trendy scriptwriters to portray murder as some sort of personality quirk. The not-so-subtle implication is that we're all murderers at heart. Therefore, we're expected to show lots of sympathy for those killers we are presented with week after mind-numbing week. One week it's a social-climbing murder, then a corporate ladder murder, then a jealous spouse murder, then a hot-headed teenage murder. The sickening list goes on and on. Burned out by all the violence? Take your spouse out to a fun, relaxing dinner murder theater!

Movies

Thanks to a much more prominent rating system, movies have been able to ramp up the violence level without much public outcry. Kids aren't allowed to see R and X rated films, right? Not really. If they can't catch adult films on cable television at home, kids can always go to a multiplex theater and simply wander into any movie they choose to, regardless of its rating. Ushers are an expense most theaters can't be bothered with today.

Ever since George Lucas swept us all away with "Star Wars" over two decades ago, computerized special effects have revolutionized the big screen. At long last movie images can match the wildest writer's imagination. The trade off here is now the number of opposing warriors slaughtering each other has risen from a few hundred real actors to tens of thousands of computer-generated ones. Bigger and bloodier is very in.

Now the trend to computerized images has spread to include *all* the actors in a movie, not just the faceless extras. This disturbing trend began with big name movies in 2007. "Beowulf was the notable example of replacing all actors real images with computerized versions of themselves. Hollywood seems obsessed with draining movies of their humanity. Are superstar actors the next endangered species?

If a culture is remembered by the quality of the stories it tells, what do modern movies say about America? Take away the violence, screaming, and dazzling special effects, and what's left? Usually a shallow, selfish tale of a few lost souls trying to survive in a cold, crazy world. Morality, inspiration, enlightenment, and depth of character—those features of great movies—are strangely absent. Will future generations of Americans gain any wisdom or inspiration by viewing all this loud, violent fare?

This very sad degradation of media art has a common thread: radical change. Corporate profits depend upon the ever-shortening attention span of the average viewer; therefore, new themes are essential to profits. New themes by definition entail breaking down old traditions and moral values. As a result, malignatech smoothly infiltrates modern storytelling. Serial killers with assorted weird masks are suddenly cool. Bodies being hacked to bloody pieces is cool, cutting edge (pardon the pun) entertainment. Morality and decency are so old fashioned. Can we even begin to calculate the destructive power of today's brutal, sleazy video games on young minds? How does identifying with crooks and killers affect a child's outlook on adult life? Why is even the religious right now unwilling to protest the media's deliberate destruction of our basic moral values?

Once again the answer seems to lie in our country's extremely naive assumption that technology is inherently good. We are desperate to trust the power that brought us out of those damp caves eons ago and into our custom-built mansion. Anything which elevates us above cold, cruel Mother Nature and her lesser animals is to be embraced without question. This futile game will soon have a bad ending. The more we separate ourselves from Nature, the more we self-destruct. Read your newspapers and magazines very carefully. Notice the shocking growth of modem ills, both mental and physical, in all age groups. Booze, drugs, smoking and gambling used to sum up addictive behavior. But the list keeps growing: girls cutting themselves, kids sniffing paint and glue, getting high from large doses of cough medicine, which can also be done with 140 non-prescription medications. [1] Add in the constantly-growing list of cyber crimes and it's easy to see how quickly rationality is going out the window. As a consumer society, we're expected to try everything, at least once. Sadly, that once is too often fatal. Do corporations or the federal government care about this deadly trend? Obviously not. They care about the only two factors running America;

money and power.

Art as a celebration of natural beauty and tranquility is becoming a thing of the past. Our warped concept of Nature as a force to be subdued and looked down upon has perhaps permanently tainted our view of art. How many poems and songs are being written about Nature? Let's be honest about the new green movement. It's essentially a cleanup game to protect one species—us—from our own insane polluting of the air, water, and ground. It's an excellent start, but is a small, belated step toward getting back into harmony with Nature. Attaining that harmony is our only chance for survival.

CHAPTER 10—"TO THE INTERNET...AND BEYOND!"

Knowledge, wisdom. These are two words we don't hear very often anymore. Data, information, numbers, stats. These are words we hear constantly. Is this contrast a problem? Now that our all-knowing internet and computers have taken over, do we really need knowledge and wisdom?

In order to answer this question, let's briefly define these two words. Knowledge is an awareness of information, as in memorizing all the American presidents. Wisdom is knowing the relationships among all your information, as in the fact that Abe Lincoln, one of our least educated presidents, was perhaps our most brilliant one. These relationships are often intricate and sometimes seem to defy logic. In short, they're not quickly learned. They require careful thought and patient reasoning, something pressing buttons and watching flashing images on a screen don't provide. Acquiring knowledge, and—especially wisdom ~ takes time and effort, two things computers do their best to eliminate. Surfing the web is exciting and colorful and certainly fast. Plus, you don't have the strain of holding a book or magazine and turning all those pages that love to stick together. The path to wisdom used to mandate a dedication to one's learning media (books, papers, blackboards). Does today's student have that dedication to a glowing screen?

An area of learning which may answer that question is computerized simulations where students practice by doing. A study conducted by the NTL Institute for Applied Behavioral Sciences in Alexandria, VA, found that students retain 75 percent of what they learn in computerized simulations, which is far more than in other learning techniques. This result tracks with the whole body learning described in the brain chapter. Major drawbacks to computerized simulations are that they require active engagement and an intense concentration often lacking in young children.

Is computerized learning the wave of the future, or will today's students get so lazy and bored with learning they reject it?

Identity Theft

Back in the old days very few thieves would consider wandering into a small town and pretending they were Jack Thompson or Nellie Smith. Local residents knew most of their neighbors by sight, so an identity thief would get thrown in jail, or worse.

In our wonderful new millennium, though, things are different. We don't know all our neighbors by sight anymore, and a person's identity isn't just a name. It's often a string of account numbers: social security, credit cards, bank accounts, etc. The ease with which those numbers can be stolen and business/government's casual attitude toward this crime is shocking. Identity theft is an extreme example of business' and government's blind faith in technology, in its inability to be used for evil purposes.

To add insult to injury, certain companies have a vested interest in *not* cracking down on identity theft! How can credit bureaus and financial institutions be so out of touch with consumers? It all started when many states began passing legislation to protect consumers against this sleazy crime. The key feature of this legislation was the long-overdue right for a consumer to activate a "credit freeze," which requires a prospective lender to first get your permission before issuing new credit in your name. Seems like common sense, right? Yet, until recently, all a thief needed was your social security number and address—both easily obtainable—to open an account in your name. A credit freeze on a person's account mandates that any company, such as a department store, wanting to issue a credit card must personally contact the person to verify pre-established private information. Any individual has the right to unfreeze his or her account to obtain new credit.

How could anyone be against such excellent consumer protection? Credit bureaus and financial institutions don't like credit freezes because they slow down their business. Money always talks! Credit bureaus are in business to sell credit information to financial institutions, who are in business to issue new credit, such as loans and credit cards. 1 Apparently, the 3.4 million new phony accounts opened each year and the average loss of almost $10 thousand per victim (usually absorbed by the businesses) are all simply considered a cost of doing business. 2

It gets even more bizarre. Industry groups in 2006 actually—I'm serious on this!—lobbied Congress in favor of an insane federal law authored by Ohio's Republican Rep. Steve LaTourette that would supersede existing state credit freeze laws. This new law would deny Americans access to credit freezes unless a person has ALREADY been a victim of identity theft!! In such a case the victim would be required to produce a police report of the crime before being allowed to put a credit freeze on his or her accounts. 3 What did I say about malignatech warping rationality out of shape!

Unlike the public and large companies, identity thieves are constantly innovating. One increasingly popular technique is targeting tax filings and returns. In 2007 our Federal Trade Commission received 20,782 complaints related to tax form theft problems. This figure represents a 150 percent jump since 2003.

In one case (where the thief was actually caught and convicted) a Florida woman used her position as a Girl Scout troop leader to get personal information about her troop's members. She used a fake medical-release form to obtain children's social security numbers. She then used that data to file electronic federal income-tax returns. She even filed returns in the names of her own children!

Another type of tax-related identity theft involves using a stolen social security number to get a job. When the thief's employer later files a Form W-2 for his or her wages, the IRS computers attribute that income to the social security number's rightful owner. When he or she gets an IRS notice asking about unreported income, the fireworks start. Instead of doing more to tighten its filing system to prevent fraudulent filing, the IRS has often assumed its ham-handed role of blaming the victim. According to IRS National Taxpayer Advocate Nina Olson, "Too often, victims of identity theft receive more scrutiny from the IRS than the perpetrators of identity theft."

Considering the damage identity theft does to both individuals' financial lives and the companies that pay actual dollar damages, you would assume any company involved in credit cards (which is almost all large businesses) would be extra vigilant about protecting such valuable consumer information. Wrong assumption! Since businesses have already factored in the $53 billion damage costs and also know it's unlikely any theft of information from them will actually be traced back to them for causing the theft, American companies have very little incentive to carefully safeguard consumers' personal data. The result has been a constant list of sometimes massive hacker attacks and losses of physical records:

- In the summer of 2005 Atlanta-based Card Systems' lax oversight of credit card transactions caused as many as *40 million* credit cards to be exposed to hackers. What makes this crime even more inviting to those hackers is the estimate of research firm Gartner Group that fewer than 1 in 700 identity thieves is ever convicted.

- An obvious weakness of our entire financial system is the presence of information brokers who sell consumer records to other companies. What happens when someone creates a phony company in order to buy sensitive consumer information? A company called Choice-Point found out when it sold such info on 145,000 consumers to a fake company. Why didn't Choice-Point thoroughly check the so-called company's credentials? Being very responsible in this day costs more money; the bottom line is sacred. Once again, malignatech has taken over the corporate mentality. Cut all costs to the bone, even vital costs. Hustling every possible buck has become the universal obsession.

- BJ's Wholesale Club provides an even worse example of this sad trend. The company left information unencrypted on computers, stored records in violation of bank-security rules, failed to use a firewall to prevent wireless intrusions, then protected the information with routine passwords which came with the system. Thieves stole information and bought millions in goods.

- Bank of America simply lost backup tapes containing information on 1.2 million government workers.

- Time Warner gave backup tapes containing almost 20 years worth of employee names and social security numbers to Iron Mountain for secure storage. The unencrypted tapes disappeared from the back of a van, exposing 600,000 people to identity theft.

- A startling new hacking scam should have every financial institution in America scared stiff. Beginning in early December of 2007 and ending in late February of 2008, 4.2 million customers of Hannaford Bros. Co., a grocery store chain, had their credit and debit card numbers stolen *as* they were being transmitted from the stores for routine purchases. Officials discovered that hackers had somehow installed identity theft software in many of the chain's stores. Once again a modern corporation got outwitted by some very sophisticated hackers. How could a large company allow such a serious breach to occur?

How many Americans are aware of the seriousness of this massive crime wave which affects 9 million of us? Identity theft has been the fastest-growing crime in this country for the seventh year in a row. Some victims spend $1,400 of their own money and 600 hours to repair their credit.

Our immediate solution to this insane problem is obvious: treat *all* financial account numbers as valuable property, to be monitored and protected with much diligence. Our broader solution is much more complicated and involves taking a hard look at the illusion of credit. Ask the average consumer why he or she uses credit cards instead of cash and the answers will be predictable: convenience, control of finances, the ability to buy it and enjoy it now, then pay later, the status of being able to flash a fancy card, etc. Convenience is a valid point, since it's easier and safer to carry a credit card than a wad of bills. But why don't Americans wake up to the fact that credit is essentially a trap for business to control you? With a savings rate dipping below zero, America has become conditioned to live using other people's money. When you do that, however, you wind up living by their rules and their interest rates.

"What's your credit score?" in all those ads really means

"our evaluation of you can control your financial future." With personal financial information slipping into the public realm, credit bureaus are dying to pry into your finances as much as possible. Blame this inclination partly on the economy. After the subprime mortgage disaster, any company involved in lending is tightening its standards. A major reason for the new attention to whether you pay your rent on time and other personal habits is the credit industry's desire to reach out and grab a new market; namely, the estimated 50 million Americans with little or no credit history. This group of young adults, senior citizens, and immigrants has inspired the creation of an alternative credit-scoring system to assess credit worthiness. 5 Do you get the impression Big Business wants *everyone* to have a credit score?

Remember that all your personal financial information is easily accessible to all those curious agencies of our federal government. As we willingly surrender our privacy to their greedy gaze, we lose even more control over our lives. That "magical" combination of credit accounts and computers to manage them has come together to bankrupt this country with a debt so enormous it can obviously never be repaid. Malignatech is brilliant at disguising itself as your salvation when in reality it's just the opposite.

Chapter 11—TRYING TO STEAL THE SUN

Despite what critics may say, a perfect nuclear power plant *can* be built. The procedure would be simple. We'll get someone else to pay for it, so it won't cost the U.S. taxpayer one cent. We'll let someone else handle all the toxic, virtually eternal waste. We'll get someone else to constantly refuel the plant so fuel costs will be zero. We'll get the world's best nuclear expert to manage the plant. We'll build it big enough to supply all the planet's energy needs. Finally, we'll site it far enough away from population centers to insure perfect safety from radiation leaks and meltdowns.

As you might have guessed by now, someone has already built this perfect nuclear power plant, millions of years ago. Look up in the sky on a clear day and you will see it, safely sited 93 million miles from Earth. It's called the sun. Every day it showers us with more than four thousand billion kilowatts of free energy!

You would think that after all these thousands of years of being civilized, mankind would have learned how to harness this incredibly abundant power source to provide most of our energy needs. No such luck. Instead, we have essentially tried to duplicate the sun with nuclear power plants. Of all the examples of malignatech, nuclear power and weapons are among the very worst. Everything about these two has been a disaster. Yes, two atomic bombs ended World War II. Yes, nuclear power provides about 20 percent of America's electricity. But the price we have paid and will continue to pay for thousands of years is way too high.

Nuclear Weapons

Our government's record of constructing and testing nuclear weapons has been an absolute disgrace! A dozen weapons facilities all over the country have released radioactive waste that could require decades to clean up at a cost of $200 billion of *your* money. 1 Early open-air tests in our deserts exposed nearby troops and the entire world to airborne radiation. From the 1940's through the 1960's, radioactive waste was stored in barrels which were often just dumped into the ocean. 2 This means for the past sixty plus years the Earth's air, land, and water have been constantly contaminated with radioactive waste from not just America but also all the other countries producing such waste.

It is critical to remember this is no ordinary waste. According to a 1983 U.S. National Academy of Sciences report, high-level radioactive waste will require 3,000,000 years to completely decay! Even low-level waste can be dangerous for 10,000 years. Where do we get the right to PERMANENTLY pollute this planet?!!

Our government began making nuclear weapons during World War II and never stopped. In our rush to outspend the former Soviet Union we cranked out thousands of these weapons of mass destruction, until we had an estimated 7,500 of them! How many bridges, schools, and houses could be built with even a small fraction of that budget?!

Even after sixty years and more than 1,000 test blasts, our scientists still have only a limited understanding of what happens inside a nuclear warhead when it explodes. They are also still trying to determine the effects of aging on nuclear warheads sitting in storage. 3 The Bush administration had wanted larger "bunker buster" warheads to go after deeply buried targets. Congress, in a surge of patriotism, authorized almost $30,000,000 to study making a new, more reliable warhead. 4

Now comes the embarrassing part. Go into any high school in America and ask a sixteen year old what he or she thinks of all the hoopla over these missiles and our plans for a "Star Wars" missile defense system capable of shooting down enemy warheads streaking toward America. He or she would very likely reply (with a smirk), "ancient history, dudes." When asked to explain, the teenager would disdainfully state, "All those missiles and even a perfect "Star Wars" system are a colossal waste of billions of bucks. Haven't you guys heard, missiles are obsolete! Every rogue nation on Earth knows we have a zillion spy satellites up there watching their every move. NO ONE is going to risk getting incinerated by us by launching a missile. Terrorists have gone low tech. Their new delivery system is a tramp steamer, a Piper Cub airplane, or a hippie van coming across an unguarded border crossing point. A lot cheaper than a missile and—big and, dudes—it *can't* be detected by any spy satellite. No missile, no retaliation. We're screwed, dudes."

At that point you quickly leave before the kid sees your red face and open mouth. Billions and billions of *your* dollars are still being spent on an obsolete weapon. Can you spell malignatech?

Nuclear Power

It would seem to be a logical progression to use our knowledge from creating nuclear weapons to build a reliable nuclear power industry. Unfortunately, logic has very little to do with nuclear power. Its one major advantage, lessening our dependence on foreign oil, is more than offset by numerous negatives.

Ever since the first nuclear power plant began operating in 1957, this industry has been a classic example of man meddling with a technology way beyond his understanding.

In the industry's early years there were no plans for disposing of nuclear waste because everyone assumed spent fuel rods could be reprocessed. The one reprocessing plant built in 1968 in West Valley, NY closed six years later. Reprocessing didn't work because the used uranium and plutonium were too contaminated to be reused. Complete reprocessing of the spent uranium made it 15 times more expensive than new material. 5 So the buildup of nuclear waste began, and continues to this very day.

Everything about this power source is outrageously expensive! Instead of testing and refining our standard design for a power plant, the industry tried to play Frank Lloyd Wright by custom designing and building each plant. 6 With 441 nuclear power plants all over the world, that's a lot of different designs. 7 Costs soared as each plant's bug caused more expenses. Shoddy work often had to be torn out and redone. Lax government oversight was the rule. The Atomic Energy Commission didn't inspect plants being built, only completed ones!! Major problems were ignored to avoid expensive—and embarrassing—rebuilding. 8 The last U.S. nuclear power plant was built in 1978. Since then orders for over 100 plants have been canceled. Multi-billion dollar plants such as Marble Hill, WA and Zimmer, OH have been built but never put into operation. Guess who pays for these gigantic white elephants? 9

Speaking of paying, did you know that another $7,000,000,000 of your tax dollars has been blown during the last ten years for cleaning up Rocky Flats, a former nuclear weapons plant in Colorado? It was the largest environmental cleanup project in U.S. history. Up until 1992 Rocky Flats made plutonium triggers for nuclear weapons on a 6,000 acre site containing 800 buildings. Plans are to open part of the site as a federal wildlife refuge, but some areas are too contaminated to be open to the public. 10 Be on the lookout for animals that glow in the dark!

Soaring costs have been the norm for the nuclear power industry for the last 50 years. Why should this industry require such massive federal subsidies? Your federal government (using *your* tax dollars) assumes liability for any nuclear accident damages over $10.5 billion. There are also new construction loan guarantees, regulatory delay financial guarantees, and massive cleanup costs all paid for by you. Why can't such a mature industry pay its own way? 11

The end result is that nuclear power now costs about twice as much as coal-generated power. 12 Despite this sobering fact, energy-hungry countries are being forced to plan many new nuclear power plants because of pollution concerns and the growing uncertainty of oil supplies. China alone plans to add as many as 30 new nuke plants by 2020. Worldwide, there are 30 new plants under construction, with 70 more being planned.

In keeping with everything nuclear, there is a major problem with building new plants; namely, we are rapidly running out of the uranium needed to create nuclear power. In addition to the 441 nuclear power plants worldwide, there are another 284 nuclear reactors used in research centers and 220 reactors used to power ships. This total of 945 reactors needs an average of 680,000 pounds of uranium a year to operate. Since the end of the Cold War some of the supply has come from recycling nuclear weapons. That supply has almost been used up, and new supplies are not being produced fast enough. In 2005 total world consumption of uranium was 200,000,000 pounds, but only 100,000,000 pounds were produced. This is a gigantic shortfall! Uranium's price has consequently soared, going from $7 a pound in 2000 to over $36 by the end of 2005. 13 The price will very likely just keep rising because of the shrinking supply. What happens when a plant runs out of uranium? Lights go out. Can't happen for many years, you say? It already happened in 2003 in Canada.

In a perfect example of malignatech logic proponents of nuclear power constantly remind us that it produces no airborne pollution: no soot, no carbon dioxide, no carbon monoxide. These people conveniently forget to mention that this extremely complicated power-generating process produces toxic waste at each of its many steps: uranium enrichment, fuel rod use, cooling water, tools and protective clothing. This is no ordinary pollution. It is the most dangerous, longest-lasting pollution ever created on this planet. It's been estimated that even a 30,000,000th of an ounce of plutonium could cause cancer if inhaled. 14 As stated earlier in this chapter, even low-level nuclear waste can be dangerous for 10,000 years. High-level radioactive waste will require 3,000,000 years to completely decay!! All of our brilliant scientists have had OVER HALF A CENTURY! to figure out a method and place to safely store this growing mound of eternal, toxic waste. THEY HAVE FAILED!

Our Energy Department recently announced that construction of the national nuclear waste repository at Yucca Mountain north of Las Vegas has been put on hold. This facility had been scheduled to open in 1998 at a cost of $60,000,000,000!! Now the Energy Department concedes that estimate may have been too low. How many schools and homes could $60 billion build? 15

Thousands and thousands of future generations will curse us for what we've been doing: *permanently* polluting this country with 44,000 tons of radioactive waste so we could watch reruns of game shows on our giant screen television sets. 16 Malignatech comes in high definition.

Nuclear power produces the most toxic waste ever seen on Earth. These power plants and their waste are, therefore, dangerous and should be carefully supervised and maintained. This industry is forever repeating its mantra: nuclear power has never killed a single American. That statement is a lie. Three operators of a small reactor in Idaho Falls were killed when their reactor exploded in 1955.17 Nuclear power's safety record is a disgrace and is actually getting worse. During the '70's and '80's alone almost 3,000 mishaps involving transported commercial nuclear materials occurred in America. That's a *lot* of mistakes! Plant accidents have happened in the U.S., Canada, England, and the Soviet Union. 18

Most of the 284 research reactors mentioned earlier are *still* unguarded, seven years after 911! Some of these reactors are still fueled with highly enriched uranium, the easiest material to use to make a nuclear bomb. Our government is aware of this shocking lapse in national security, but is so mired in red tape it can't fix the problem. In 1986 the Nuclear Regulatory Commission recognized the danger posed by these facilities and mandated that all these reactors convert to low-enriched uranium, which terrorists can't make into a bomb. The facilities were ordered to convert as soon as usable low-enriched uranium fuel became available. Our Department of Energy would fund the purchase of this safer fuel. An excellent plan which should have been carried out smoothly.

Enter malignatech! Over 20 years later there are *still* 7 reactors in this country using highly-enriched fuel that could use low-enriched uranium. What happened? The Department of Energy failed to fully fund the conversion part of the project. Amazingly, each reactor would cost at most a few million dollars to convert! With all the billions being spent on homeland security, the federal government can't afford a few million?! As a perfect clincher to this fiasco, the NRC has decided that since these research reactors have such limited funds, they are exempt from nearly all security requirements. No armed guards, no elaborate security system. 19 In other words, ignore the problem and it will magically disappear! Can you believe that government agencies you are paying billions of your tax dollars are making such incredibly stupid decisions?!

So far America has been spared a nuclear plant meltdown, such as the Soviet Union's Chernobyl disaster. That 1986 accident caused 1,000 cases of radiation sickness, 30 immediate deaths and an untold number of subsequent ones. We have, however, had a number of serious accidents:

- In 1956 construction began on the Enrico Fermi Atomic Power Plant, about 30 miles south of Detroit. The plant was built *directly* under the flight path of planes landing at the Grosse Ile Naval Air Station. Dozens of Navy planes would fly over the plant daily. In 1966, as the $120 million plant was becoming fully operational, it experienced a partial meltdown of fuel rods. It took almost a full year to determine the cause: a round piece of metal the size of a beer can that had not even been on the blueprints. 19 After the plant was shut down as a lost cause, another nuclear power plant was started nearby. At the end of 1974 work on that half-completed plant came to a stop when financing for the project dried up. 20
- In 1955 an experimental breeder reactor in Idaho Falls, Idaho, experienced a partial meltdown, and avoided a full meltdown by mere seconds. As previously mentioned, three operators died. 21
- By 1975 the major print media (*Newsweek*, the *Wall Street Journal*, the *New York Times*, etc.) revealed serious problems with nuclear power: there was no protection for nuclear plants against earthquakes; thousands of kilograms of plutonium were missing; almost half the reactors in America had been shut down for various problems. 22
- Also in 1975, the nuclear power plant in Brown's Ferry, AL, suffered a major fire which destroyed most of the reactor's electrical system. Cost to

repair the damage: over $300 million! 23

- In the infamous 1979 Three Mile Island incident in Pennsylvania over 40,000 people were evacuated. Total costs of the accident were approximately $1 billion!

- Way back in 1982 the Sandia National Laboratory in New Mexico issued a report stating that a worst-case nuclear plant accident in America could cause as many as 140,000 deaths and cost $100,000,000,000 (yes, billion!) in damages. And backers of nuclear power *still* want to build more of these ticking nuclear bombs!

- In 1991 an on-site emergency had to be declared at the Nine Mile Point 2 reactor in New York when power to key instruments used in monitoring the reactor was lost.

- By now it's obvious to most people that nuclear power plants are probably the most complicated operations in the industrial world. Control rooms have hundreds of gauges and require careful, highly skilled operators to monitor them. It should be of *major* concern that since 1999 six nuclear plant operators in this country have been fired for substance abuse. The latest incident occurred in 2005 when a control room *manager*—not just an operator—at the Oyster Creek nuclear plant in New Jersey tested positive for cocaine. 23 Does it bother you that a drug addict has the power to cause an accident which could kill 140,000 Americans?!

- In 2006 at the La Salle County nuclear plant south of Chicago three control rods used to shut down the reactor were not fully inserted, resulting in a complete reactor shutdown to avoid having nuclear fuel escaping and contaminating the cooling water. 24

- Research into developing nuclear fusion (as in sun power) has been slow and underfunded. Fusion produces little radiation waste; it presents no danger of explosion or meltdown; its basic fuel, deuterium, is found in ocean waters, so the supply is essentially unlimited.

Since nuclear power plants must be built fairly close to cities to minimize the cost of transmitting power over expensive power lines, the problem of how to evacuate a large urban area quickly becomes extremely critical.

Our experience with Hurricane Katrina should be a sobering reminder of how shockingly incompetent our local, state, and federal agencies are at disaster management. This costliest disaster in our country's history killed more than 1,300 and caused damages that could total over $100 billion. Eighty percent of New Orleans was flooded. Most of this flooding was caused by more than 50 breaches in drainage canal levees built by the U.S. Army Corps of Engineers. 25 All this happened in spite of a supposedly-prepared Homeland Security Agency and a couple of days warning.

Compare this disaster to a nuclear accident. Instead of two days' warning, residents near a nuclear plant could get two minutes' warning. If the result was ten hour traffic jams, as happened in Hurricane Katrina, escaping the far-reaching radiation would be out of the question. Remember that we know from the Chernobyl disaster that radiation can travel over 300 miles. This means there is *no* spot in the entire eastern United States that is not vulnerable to radiation from a nuclear plant explosion. Add in the sad fact that most cities near nuclear plants have only vague evacuation plans which often rely on untrained volunteers.

Did we learn anything meaningful from Chernobyl, which happened over 20 years ago? Damage from this world's worst nuclear disaster was staggering. Radioactivity caused 31 people to die within the first 2 months. The U.N. Health Agency estimates the final death toll will be approximately 9,300 people. Considering the fact that 5 million people live in the 77,200 square mile contaminated area, that 9,300 estimate could be very low. Over 300,000 people *were permanently* evacuated. Where did they go?

With all the close calls the nuclear power industry has had in the last 50 years, it's only a matter of time before a plant finally explodes. Earthquakes, plane crashes, sabotage by drugged-out operators, or terrorists could easily cause such a disaster. Terrorists could take over our poorly-guarded plants and cause a meltdown before any security force could react. Why would terrorists bother trying to smuggle a nuclear bomb into a major city when the bombs are already sitting nearby, waiting to be detonated?! In case a terrorist group did want to make their own bomb, the plutonium needed is everywhere: poorly guarded storage facilities, unguarded research reactors, material being constantly transported across our country, or from a nuclear power plant in their own country. Ask Uncle Sam for some help in starting a nuclear program, and we'll be glad to help you destroy us. Our government is doing exactly that right now by helping our "sometimes" ally, Turkey, start its own nuclear power industry. Add nuclear power to a dictatorial government and the result is almost always nuclear weapons. The malignatech disease has become a pandemic.

Instead of pumping untold billions of taxpayer dollars into the insanity of nuclear power for the last 50 years, our federal government should have been focusing on the basis of our power industry; namely, our creaky national electric grid. Even though the states control their own power generation and transmission lines, federal assistance is badly needed.

A recent study by the journal Physical Review E discovered that of the over 14,000 substations which generate and distribute energy throughout America, some are much more overloaded than others. Our deregulated power industry now has financial incentives to shift power to the areas where the highest rates can be charged, leading to overloaded power lines. This study determined that "During a cascading failure, in which the high load substations fail in sequence, the model shows that the loss of only 2 percent of the nodes (substations) causes a catastrophic failure of the entire system."

This sort of widespread blackout has already happened on August 14, 2003. On that warm summer day 50,000,000 people in the Northeast, Midwest, and parts of Canada lost all power. Prior warnings from specialists in the field and a government report in 2001 on the need for improvements in the grid had not been heeded. Cynics are claiming the federal government didn't mind this sort of accident because it strengthened their case for a federally-owned national grid. In the uncoordinated power system we now have, states haven't had the money to invest in new transmission lines, but they still want to maintain control of their own systems.

We're talking big bucks here! Just to *maintain* the country's 157,000 miles of high-voltage transmission lines could cost $56,000,000,000 over the next decade, according to industry consultant Eric Hirst. Modernizing the grid with devices such as digital outage arrestors and super-conducting wire could cost twice that amount.

The most damning evidence against nuclear power is actually the industry's own opinion of itself. Back when it began in the 1950's the industry knew it couldn't afford to insure its power plants against a major accident. The insurance industry also knew this dirty little secret, so it offered only token coverage. Naturally, the nuclear power industry took the only course left, con Uncle Sam into providing the coverage. The resulting 1957 Price-Anderson Act provided only $500 million federal coverage for an "extraordinary nuclear occurrence" (read "nuclear explosion") at a power plant. 26 What other industry requires the federal government to insure them?! That ridiculous coverage amount was recently increased to $10,800,000,000 per accident, which is still ridiculous. 27 The 1982 Sandia National Laboratory report estimated a nuclear plant explosion could cost $100 billion. Who pays the extra $89 billion? YOU DO!! Anything over that small cap is insured by *your* tax dollars. In other words, if you and your family have to evacuate your $500,000 home in a nuclear accident and buy a new home far away for at least the next few decades, you might get a small settlement check. You might also get nothing. Your home and most of your possessions are now radioactive, your life and health have been seriously endangered, but the insane industry which ruined your life is not legally required to pay you one cent! Think carefully about all this if you live within 300 miles of a nuclear power plant.

In summary, it should now be obvious why nuclear power is Exhibit A for malignatech. This form of energy is the most TOXIC, DANGEROUS, EXPENSIVE, and UNRELIABLE ever created! Everything about it is an insult to Nature and common sense.

There is only one sun in this solar system. Until we unlock the secrets of fusion power, we should keep anything nuclear 93,000,000 miles away, where it belongs.

Chapter 12—WHEN LARGE PIGS FLY...YOU GET MASSIVE PORK IN THE SKY

Many Americans complain about earmarks in congressional bills because of the insane pork barrel projects they represent. That Alaskan "bridge to nowhere" was one of the most notable examples of governmental waste. Yet few Americans seem to realize our own NASA has degenerated into this country's most expensive port barrel project.

Our Fantasy's Beginning

Since ancient times when the mythical Icarus strapped on wings and flew too close to the sun, man has dreamed of space flight to explore other worlds. The all-important nudge this fantasy needed to become reality occurred in World War IIs rocketry competition between Germany and America. Luckily for us, many German rocket scientists moved to this country after the war, when the Cold War between the West and the Soviet Union gave an even bigger boost to space flight. As the two sides vied to create the most powerful intercontinental ballistic missiles (ICBM's), the space race took on the aura of a war in the sky.

Then came Sputnik, the world's first satellite, on October 4, 1957. It was a gigantic coup for the Soviet Union. America responded quickly to this perceived military threat, and the next year President Eisenhower signed an act creating NASA. Developing and launching communication satellites were the space agency's initial focus, with AT&T's two Telstar satellites being probably its most famous launches.

But NASA was looking beyond our atmosphere at a manned moon landing. Even in the early '60's there were critics of manned space flight. One top scientist flatly stated it had "no military justification." 1

In April, 1961 the Soviet Union delivered another major blow to our national ego when Yuri Gagarin became the first human to fly into space, with a two-hour flight over his country. President Kennedy vowed to win the space race by being first to land a man on our moon.

One month later Alan Shepard rode a rocket 115 miles above Earth to become the first American in space. The space race became a political/military circus, as both countries vied for the biggest newspaper headlines. We were beaten a third time when Valentina Tereshkova became the first woman in space.

After Apollo

Our Apollo moon missions won the space race and established America as top dog in space. With the exception of a space capsule fire which killed 3 astronauts, NASA's Apollo work was brilliant. After that program was shut down in 1973, however, the space agency began to fall apart.

With no more moon "mountains" to conquer, NASA had to scramble to create a new mission or face a permanent vacation. President Nixon knew he couldn't completely halt manned space flight missions without major political consequences, so he agreed to the shuttle program as a practical continuation. It wouldn't be as expensive as Apollo and would have more military uses. At the planning stages the Pentagon wanted everything to do with our shuttle program to be classified. NASA refused, knowing this would ruin its reputation for openness. 2

Instead of an Alaskan "bridge to nowhere," NASA created a "space station to nowhere." When Skylab went into orbit in 1973, its supposed purpose was to study space's long term effects on humans in anticipation of bigger missions, such as flights to Mars. The fact that no such flights were planned didn't seem to matter.

Political opposition to the shuttle program's cost began to grow. Remember, this was over 30 years ago when a billion dollars was still a lot of money. From an initial budget of $1.2 billion in 1962 for the entire agency, NASA's budget for just the shuttle program in 1971 grew to $5.1 billion. NASA had requested twice as much. The reduced budget led to some dangerous design compromises, such as solid fuel propulsion and no escape modules for the crew.

With no real mission other than to keep NASA busy enough to stay in existence, our shuttle program had the odds stacked against it from its very beginning. There was always intense pressure to make the program cost effective. By the late 1970's, though, the military's "black' space program could ignore budget constraints, as it spent more than NASA did on its civilian program. Spy satellites became powerful enough to read license plates on parked cars. As constant budget cuts forced staff reductions, morale sagged.

Design problems plagued the shuttle program from the beginning. Engine design mistakes caused engines to melt down and explode. Columbia, the first shuttle craft, required 31,000 ceramic tiles to cover its hull! Why was such an ancient, unreliable, and expensive material used on the world's most expensive aircraft?! It gets even more insane! As part of a good-citizen program, Rockwell, Columbia's manufacturer, hired ex-convicts to install the tiles. Inspectors found puncture holes on the tiles, made by the ex-convicts to extend their jobs!! NASA then decided to save money by hiring college students to do the tile installation. Unbelievable!

By Reagan's election in 1980 the military was firmly entrenched in the shuttle program, doing most of its work inspections. Under Reagan the sense of pride and accountability at NASA deteriorated. Professional men like Chris Kraft resigned in disgust. The administration's "Star Wars" mentality further militarized the shuttle program. Cost overruns became the norm, as politics overwhelmed the space agency. It became a pawn of the military, with budget funds being diverted for pet projects like the B-1 bomber. Revolving door jobs meant contractor oversight was greatly weakened.

Costs soared. Each of the five shuttle craft was estimated in 1972 to cost $675 million. By 1983 the real cost of each had rocketed to $ 1.47 billion, over twice as much. Payload costs had been estimated to be in the $100--$270 per pound range. By 1983 *each pound* orbited cost more than $5,000!! NASA's plan that the shuttle could pay for itself turned out to be a wild fantasy.

Fatal Fantasy

Fantasies are alright as long as they don't harm anyone. Unfortunately, the space station's only clear purpose was to give the shuttle program a reason for being. As such, the space station became a fatal fantasy. Two multi-billion dollar spacecraft and 14 lives were destroyed by this pork barrel fiasco.

Challenger

There seems to have been a cosmic irony in this 1986 disaster, when a non-moving, low tech part caused the destruction of man's most complicated machine. Burning of the O ring seals had been an ongoing problem, but nothing was done to address it. Cold temperatures the night before launch made the seals brittle, but once again the problem was ignored. When Challenger exploded at 65,000 feet soon after launching, there was no escape module to save the 7 astronauts aboard. Observe the logic of malignatech: multi-million dollar jet fighters have escape modules to save one pilot, but multi-billion dollar space shuttles can't afford such life-saving devices for a crew of 7.

Columbia

As shocking as Challenger's destruction was, the 2003 crash of Columbia was an *inexcusable* crime for which many people should have gone to prison for extreme criminal negligence!

Once again, a serious design flaw was built into the shuttle's giant fuel tank stacked above it for launching. Foam insulation covering this tank was essential to keeping the fuel chilled to as low as 423 degrees below zero. NASA was well aware of the danger of foam insulation breaking off the fuel tank and striking the shuttle because such an "accident" had happened during EVERY shuttle launch!! There had been 111 launches! Since 1981 shuttle craft had returned from their missions with an average of 143 damaged tiles! Nothing was done to stop this extremely risky syndrome, even though the space agency had a firm rule against allowing anything to strike the shuttle. NASA managers considered foam debris striking the shuttle an "accepted risk," even though they knew such debris could cause catastrophic damage.

During Columbia's final 16 day mission officials suspected there might have been serious damage to the shuttle but decided *not* to photograph the shuttle in space to check for sure. A piece of foam the size of a briefcase but weighing only 24 ounces had damaged the left wing's underside, causing it to fail on re-entry. 3

There were striking parallels between the Challenger and Columbia disasters: engineers' concerns ignored, increased acceptance of deadly risk, poor communications, and schedule pressures. Considering that the entire program was a pork barrel concoction, why were there such schedule pressures?! These two shuttle disasters happened 17 years apart. During that time the consensus at NASA was obviously that if they ignored the foam debris launch problem long enough, it would simply go away. Such a combination of colossal stupidity, arrogance, and criminal neglect defies belief!!

Nonstop Waste

As if the shuttle program weren't a big enough waste of *your* tax dollars, consider NASA's recent fiasco of spending $700 million on a 9.5 year mission to study Pluto. Launched in 2006, the rocket contained 24 pounds of plutonium for power. Can you find *any* logic to this latest insult to our budget? Pollute space with a big hunk of plutonium to study a mass of ice which is 3.5 billion—yes, billion—miles from Earth! How many roads, bridges, schools, and houses could $700 million buy?! 4 Malignatech has obviously taken control of mission control.

Like any normal failing federal agency, NASA will not admit its days are numbered. America is broke, so trying crazy gambits like a new $104 billion moon mission is a sure sign NASA's brain is still lost in space. Private spaceflight programs are showing genuine progress, with at least a dozen companies working on projects. 5

Wouldn't it be a great change to take our space agency's $16 billion annual budget and actually use it for worthwhile projects on Earth! Look for water in the parched Southwest instead of on Mars—radical! Repair our disintegrating power grid, water and sewer systems, roads and bridges, health care system, and low income housing instead of sending billion dollar rockets to planets we will *never* set foot on—radical!

Many ivory tower scientists would view my suggestions as unenlightened ravings. Their mantra cracks me up: "We must spend whatever it takes to colonize other planets to preserve the human species." Translation: "We've trashed Earth so badly and will probably destroy it someday soon with a nuclear war, so we desperately need to find some other planets to trash."

Urgent memo to all scientists on this planet: Earth is *all* we have! No other planet in this solar system is inhabitable. Plus, the nearest solar system to ours is several light years away. If you deep thinkers want to use your *own* billions to build a fancy rocket and set out for Alpha Gamma Slamma, please feel free. If you find any intelligent life out there, please send them back to Washington, D.C.

Chapter 13—IT'S 1984 FROM NOW ON

Those of you who have read George Orwell's chilling book, 1984, will remember the basic scenario: a lonely hero caught in a country constantly at war with some nebulous enemy. Sound familiar? It should, because ever since World War II America has become that war-obsessed country Orwell warned us about. Why do such peace-loving people have such a militant government? Why has our government gone out of its way to start so many wars of aggression? The only two times our country has been directly attacked, Pearl Harbor and the World Trade Center, had *obvious* federal government involvement. If that sounds like too much conspiracy theory, study the details of both attacks and you'll discover enough manipulation of our enemies and security lapses to fill a book.

There are a number of compelling reasons why war has become America's new growth industry. All of them are a sad commentary on what used to be the world's greatest champion of freedom.

Democracy Is Fragile

Most experiments in letting people rule themselves have not been great successes. Ever since ancient Rome, a power-hungry elite has always been ready to pounce when the masses become too lazy to live up to their key responsibilities of supporting themselves financially and guarding their rights. The result is a dictatorship (dream on, George) and some form of socialist government.

Malignatech has accelerated the destruction of our democracy with a double whammy. First, this disease has weakened and distracted our masses with every labor-saving device imaginable and a deluge of info-tainment devices. Then, while we have been desperately searching for the perfect giant screen HD television entertainment center, our restless federal government has been busily blowing tens of billions of *your* tax dollars on increasingly complex, expensive, and lethal weapons systems. Why do we need such a vast array of super tanks, planes with giant lasers (more on this later), and stealth fighters? Our country's location makes us one of the most secure countries on Earth. Oceans on both coasts, a narrow land connection to South America, and a vast forest to our north have us extremely well protected. Your federal government needs these powerful weapons to fight wars of aggression and keep all those hungry military contractors in business.

Starting a "War on Terror" gives the federal government a perfect excuse to start restricting our rights. After the 911 attacks Bin Ladin family members were secretly flown out of the country (while all other air traffic was grounded), then five year olds started being searched at airports! Please explain that. These restrictions start out as temporary war measures then become permanent and more numerous until they become an all-out assault on our constitutional rights. What are our congressmen and women and senators doing to safeguard our precious rights? Basically, very little. They're too busy developing pork barrel projects to get themselves reelected. Meanwhile, the feds continue to dream up more invasions of our privacy, restrictions on our finances, and vague excuses for terror alerts.

Why don't more Americans see through this socialistic scam? Partly because they're too busy seeking empowerment and escape in Cyberland, with its violent video games, endless porno websites, and mindless chat rooms. Probably the main reason is the Bush Administration's clever flag waving tactics which have conned most of us into believing that wartime necessitates losing major freedoms.

Malignatech's presence has been most shocking lately in the use of torture on terrorist suspects. Since we have the biggest weapons and most advanced technology, the malignatech mindset has taken over our tactics in Iraq. Ignoring local customs, murdering civilians, torturing and murdering prisoners have become standard operating procedure. We have betrayed our own ideals of justice and disgraced our country in the eyes of the civilized world. We supposedly invaded Iraq to bring democracy to a war-torn country. Why would any people want this brand of power-hungry control? Torturing prisoners is also pointless because pain will make anyone admit to anything. History will not be kind to that lying frat boy who plunged us into this country's most disastrous war.

Our Obsession With Failure

America is a bold, pragmatic country that worships success. Why, then, have we spent the last half century getting swept up in wars we either lost or couldn't win? Why can't we learn the lessons of history?

Did we need to get involved in Korea? Why didn't we see Vietnam for the colossal scam it was *before* we let LBJ start that conflict? Why don't most Americans see the Iraqi War for what it is, another Vietnam whose sole purpose is to guard our oil interests? George Bush blatantly lied about there being weapons of mass destruction and terrorists in Iraq to justify our invasion of that mess of a country. Why do we pay our intelligence bureaus billions of dollars a year so they can lie to us when it suits their purposes? Ironically enough, Osama bin Ladin provided us with more accurate intelligence than our own government when he predicted that America would invade an oil-rich country in the Middle East.

Our shameful history of involvement in this region has

always been based on one commodity, oil. Since America runs on it and big business controls our government, we have spent the last 50 years doing anything necessary to insure a steady supply of oil. Our most effective technique has been to prop up dictators with generous foreign aid and arms sales. In the past seven years Pakistan's Musharraf has received $10 billion worth of U.S. aid and arms. We're also very supportive of the Saudi royal family and Mubarak in Egypt. 1 How are the oppressed masses in these countries expected to view our support of their dictators? Will they begin the hate us for meddling in their lives? Why are we constantly creating converts for the terrorists?

Once again, malignatech is at work as our government ruthlessly pursues the steadily-shrinking supply of this soon-to-be-obsolete energy source. What has malignatech cost this country in the 5 years of our occupation of Iraq? The losses are staggering:

- Over 4,000 American soldiers have been killed, including 105 women
- 60,000 Americans have been wounded or contracted a serious disease
- 151,000 (estimated) Iraqis have been killed
- Several million (estimated) Iraqis have been wounded
- More than 4 million Iraqis displaced, half fleeing the country

Annual cost of this insane war is $171 billion! Included here is the Pentagon's recent replacement of 18,000 humvees with monster trucks known as mine-resistant ambush-protected vehicles (MRAP's). The humvees cost $150,000 each. MRAP's are a bit more expensive at $800,000 each!!

1.6 million American troops have served in Iraq, more than a third of them for two or more tours of duty.

After two or three tours of duty in that hellhole of a country, how are our troops expected to have any morale left? What are they expected to think when they learn that private contractors over there, who are not accountable to Iraqi laws and face far less danger, are paid over twice as much? The Bush Administration conveniently ignores all collateral damage done to American troops' families. Children of soldiers killed in action will NEVER recover from such a tragic loss. Soldiers who are parents should NEVER be sent into combat!! I know something about this subject, having spent a year in Vietnam with the Army.

Of the 780,000 troops discharged so far, one-third have been treated at veteran's hospitals and clinics, almost 200,000 for mental health conditions. 2 How will the Veterans Administration treat those wounded soldiers for the rest of their lives? Probably with its usual mix of neglect and constantly-reduced benefits. We will still be obligated to care for those shattered lives for many decades to come. For them the Iraqi War will never end. Our military's equipment has been hammered. Vehicles and weapons are being used up at six to ten times the peacetime rate. Tactical aircraft are now the oldest since World War II, 24 years old on average. In their desperate search for "cannon fodder," our military has lowered standards for education, physical fitness, and even criminal records. Damage to America's prestige has been enormous, especially among countries in the Middle East. Our sham "international coalition" is falling apart, leaving us to fight most of this war ourselves. Where do we get the right to invade a seething mass of sectarian strife calling itself a country, then attempt to impose our form of government on these people? To keep order in Iraq, we have been forced to become the new Saddam Hussein, with torture of prisoners, imprisonment without trial, accidental and intentional killing of civilians, and support of corrupt officials. All of this primitive violence degrades every American. If we truly cared about spreading freedom, we would have invaded North Korea, Myanmar, and half of Africa by now. But those countries aren't rich in oil, which is (to repeat) our *only* reason for being in Iraq.

This endless war has not only destabilized the Middle East but also fanned the flames or war worldwide. If the world's greatest superpower demonstrates that war is the new diplomacy, the rest of the world will obviously take note. Why do we compile a list of terrorist states, then provide almost *all* of them with military aid?! Those peaceful '90's have given way to a violent new millennium. By the year 2000 over 100 new wars had broken out, killing more than five million people. 3 All of which makes for great arms sales, but drags humanity back to the Dark Ages. Malignatech is hard at work destroying humanity.

Here's one final thought on Iraq. Remember when Saddam Hussein invaded Kuwait in the summer of 1990? President George Bush senior took decisive action and drove the tyrant out of Kuwait and back to Iraq. President Bush's considerable experience in foreign affairs told him not to follow Saddam all the way back with an invasion of Iraq. He obviously knew deposing such a powerful dictator would destabilize his country and very likely the entire region. Along comes Bush junior, with almost *zero* experience in foreign affairs, and does just the opposite. What had changed between 1990 and 2003? Saddam Hussein had stopped his wars of aggression and still had no known ties to terrorists. What had changed was the oil situation. It was becoming scarcer; Russia and China were looking for new supplies; the Saudi royal family was under growing threat of being overthrown by dissidents; Saddam was secretly selling oil to other countries. In short, our access to Middle Eastern oil was facing more and more restrictions.

What's the simplest solution to such a serious problem? Make up scary lies to get rid of Saddam Hussein, then invade his country. All this to guard our access to an obsolete energy source. More malignatech: greed, lust for power, and the irrational concept that oil is still a viable fuel source. Clinging to this concept could easily bankrupt our country.

War's "Glowing" Future

With America's manufacturing jobs being steadily sent out of the country, our industry needs work. Solution? Let's make more weapons to keep those terrorists pinned down over there in the desert where they can't harm any civilians. War's future as our new growth industry looks assured. Here are two of your government's more unusual military projects.

Science fiction's Death Ray In The Sky is about to become reality. After 5 years in development and a cost of $200 million each, the Air Force has created a 20 ton chemical laser which can be installed in a C-130 gunship. Even flying at 300 mph and 10,000 feet in the air, this incredible beam can hit a moving target a few inches wide with enough energy to heat the target to thousands of degrees, burning a hole in anything on the ground. Plus, the laser can do this from five miles away with a burst lasting only a few seconds. 4

Devices like this highlight how incredibly powerful technology is becoming. Try to imagine how many computer calculations are needed to enable a giant weapon traveling at 300 mph to hit a tiny target which is also moving, all from 5 miles away! Such power makes for a terrifying weapon whose obvious role is offensive, since we don't have much risk of a tank invasion from Mexico or Canada. A flying laser gun could also be extremely effective at controlling a group such as a militia or even just one house of someone the government didn't want stirring up 'trouble."

Lest you think our federal government would never attack innocent civilians, I have some disturbing news for you. For more than 30 years *your* federal government has been conducting secret biowarfare attacks on western towns. Congressional hearings in 1975, 1977, and 1994 have confirmed that the Department of Defense has been secretly testing chemical and biological agents on American and Canadian civilians. How many times in the last 30 years have you heard about this on the mainstream news?! Why has this murderous insanity been allowed to continue for all these years? Why has no one been arrested and sentenced to life in prison for these hideous, Nazi-like crimes?!

As recently as 1994 U.S. military aircraft conducted a test on the town of Oakville, Washington. Soon after a gel substance was dropped on the town, everyone became sick with flu and pneumonia-like symptoms. People were hospitalized and remained sick for months. Pets and farm animals died. When tested, the gel substance was found to contain human blood cells and biowarfare bacteria. Government agents soon came snooping around the town, asking questions and intimidating anyone who had been interviewed on television.

In 1997 Caldvvell, Idaho was the target of a dark, fibrous material falling from the sky. Seven healthy people died in their sleep when their lungs collapsed.

Between 1988 and 1999 over 30 incidents of waste dropped from low-flying planes have been investigated. Many Americans have reported seeing or feeling chemical mists falling from the sky after planes have passed overhead. Nazi Germany is still alive, here in America. *Your* tax dollars are funding these malignatech attacks on our civilian population. The Department of Defense has set aside a lot of those dollars—$10 billion—for "terrorism prevention." *Your* federal government has a network of over 100 laboratories across the country where military and contract personnel develop new strategies and devices for killing people. 5 In

1996 Dr. Leonard Horowitz stated in his book, Emerging Viruses, that both AIDS and the Marburg-Ebola complex were created in U.S. biowarfare labs.

After reading all this chilling evidence, you should realize how malignatech has warped the nature of war into an offensive monster whose ravenous appetite must be fed by the increasingly insane military-industrial complex. This is where malignatech is becoming painfully obvious. Seriously flawed weapons are being rammed into the pipeline:

- Deepwater, a $24 billion Coast Guard modernization program, almost ran aground because of poor governmental supervision. Two contractors, Northrup and Lockeed, were given authority in 2002 to oversee this gigantic replacement of the Coast Guard's fleet. Original plans called for 91 new cutters, 124 smaller patrol boats, 244 new or modernized aircraft, and new surveillance and communications systems. Costs ballooned by billions. The two contractors wasted $100 million when their design to lengthen 8 patrol boats' hulls from 110 feet to 123 feet made the hulls too weak to be put into service. This failure brought into question the two contractors' ability to build a new 418 foot cutter. In June, 2007 Coast Guard Commandant Adm. Thad Allen took back control of the entire Deepwater program.

  Why was the Coast Guard so lazy or stupid to assume two contractors could be trusted to oversee a $24 billion program?! With that amount of money on the line, the temptation to cut corners is obviously huge. Here we are back at the three R's, with no one wanting to assume Responsibility.

- A shocking example of theory replacing reality has hit the Marine Corps. Their new tilt-rotor aircraft is definite proof that rationality is becoming obsolete!

  For decades the Corps has dreamed of being able to insert troops faster into remote areas, such as the mountainous terrain of Afghanistan. Thirty years ago plans were developed for a plane which could fly faster than a helicopter, then convert to one for landing by tilting its engines vertically. Sounds like a winner in theory, but trying to combine the two types of aircraft has proven unworkable. The Osprey's many design flaws make it a flying disaster: it's three times as heavy as the CH-53 heavy lift chopper it is supposed to replace; it's only half as reliable; its cabin is only half as big; it's not as maneuverable as a chopper; it vibrates excessively; its software frequently malfunctions, leading to crashes. Our hawkish Vice President Cheney canceled the Osprey back when he was Secretary of Defense. It has failed 2 formal Operational Evaluations, most recently in 2005. Could part of the problem be that the bulky engines and propellers are mounted on the ends of the wings, their weakest part?!

  Despite all this damning evidence, the V-22 Osprey was *still* declared ready for full-scale production!! The military has ordered 400 of them! 6 This is an incredible waste of your tax dollars and—more importantly ~ the lives of your sons who may wind up crashing in one of these flying disasters.

As military weapons systems get more elaborate and expensive, they seem to take on a life of their own—regardless of the need for them or their actual ability to perform as designed. With billions of dollars at stake, defense contractors lose sight of reality and embrace malignatech like a long-lost lover. The result is the bloated monster known as the military-industrial complex, whose mission seems to have shifted from defending America to starting a war whenever our economy falters and needs a boost/distraction.

Few people were as well acquainted with this powerful hybrid branch of our government as our 34th president, Dwight Eisenhower. Of all U.S. presidents he had a unique view of both sides of our military. He served brilliantly as Supreme Allied Commander of the war in Europe, then presided over the resulting peace as two-term president. Way back in 1960, 3 days before leaving office, Eisenhower made an interesting statement in his Farewell Speech: "In the councils of government, we must guard against the acquisition of unwarranted influence, whether sought or unsought, by the military-industrial complex. The potential for the disastrous rise of misplaced power exists and will persist."

Will the American people have the wisdom to finally heed Ike's warning and regain control over our military?

Chapter 14—TAMING THE MONSTER

Technology is a fascinating illusion. It emerges from its glittering bottle looking exactly like a magical genie. On closer examination, we often discover it's actually a monster. Since it's much too big and powerful to jam back into that bottle, we're going to have to learn to tame it *before* it destroys us.

This book does not condemn all technology. We need much of it: our houses and emergency rooms and basic tools and transportation. Malignatech, the destructive essence of technology, however, is clever at disguising itself as a necessity:

- Movies and Television—were and still are viewed as marvels of entertainment. Then came competition for viewers and the cascading violence and sex we now view.
- Nuclear Weapons — developed to end the worst war in history, have now proliferated into the hands of insane dictators, becoming a permanent threat to world peace.
- Biological Weapons—are ostensibly created by governments to protect their citizens against attacking countries. All too often, these hideous weapons are used against a country's own citizens. If this crime can happen in America, you can be assured it can happen anywhere.
- Pharmaceutical Drugs—came into major production after World War II to cure major diseases. They had some success, but have since

helped spawn dozens of new diseases, while assaulting us with countless toxic side effects and outrageously high prices.

- Computers—came on the scene as the greatest technological marvel of all time. Then recently, they became Exhibit A for malaxiom #1: "Technology's powers will always be used for evil purposes to the fullest possible extent." Computer's power and anonymity have brought the nut cases out of the woodwork. Hackers, identity thieves, cyberstalkers, cyberbullies, violent video games, and elaborate viruses have turned the Internet into a primitive jungle.

- Cellphones—have been around (in theory) since the '60's, when Captain Kirk of Star Trek first used one to order Scotty to "beam me up." You'll remember Captain Kirk never put his phone to his ear, probably because he knew the device's radiation could do serious harm to his brain. Hopefully, it won't be a few more centuries before most Americans learn this deadly truth. When this marvel of compact communication was first developed, the manufacturers succeeded in convincing the federal government that cellphones were perfectly safe and didn't need testing. This was a giant scam because the radiation from cellphones and especially from their relay towers is harmful to the human brain. The younger the brain, the more damaging the radiation. In a few years all our cheap talk will very likely become very expensive when a mysterious epidemic of brain disorders sweeps the nation.

- Toxic Chemicals—are used in or created by the manufacture of far too many products made in America. As a result, our entire environment—

land, water, and air—is now polluted with 10 million manmade chemicals! The heavy metals, mercury, lead, and cadmium, are especially damaging to our bodies. Do we care about the damage this toxic mess is doing to other species, most of which are smaller and much more delicate than us? What gives man the right to harm the environment *of every* other creature on Earth?!

- Genetic Modification—has been touted as another miracle of science. Stop and think about the logic here. The human genetic sequence required hundreds of thousands of years to evolve. Modern man is arrogant enough to assume he can just reshuffle our genes like a deck of cards, without any resultant problems. Sorry, Nature—of which we are still a part—doesn't work so simply. Supercomputers may be super fast, but there isn't a brain in the whole lot of them.

All-Out Assault On The Individual

How can malignatech be such a serious threat to civilization? Can a lot of technology ever be bad?

If you live in the first world, your life could definitely depend upon understanding the all-out assault being waged by malignatech against your *entire being.* Let's review the various fronts of this battle:

1) Physical—the most obvious problem area. We have totally polluted the Earth with our 10 million manmade chemicals. Approximately 300 proven harmful substances are found in most human blood samples today. 1 Add in constant

new diseases and toxic, often useless drugs. Food laced with fat and sugar, our myriad labor saving devices, and our increasingly fast pace of living are taking a terrible toll. Diabetes, obesity, abuse of all drugs (legal and illegal), and reckless extreme sports are expanding the attack on our health.

2) Mental ~ our bodies were designed as rugged, powerful, living machines needing daily use. Modern America's sedentary lifestyle is leading to more than obesity and diabetes. Reduced lung power and all of our polluted air are cutting back on oxygen to our brains, resulting in brain fatigue. Mix in mindless media with its endless barrage of ads and you get a stressed-out society. Our murder rate may be down but our violence level is trending upward: road rage, war, school shootings, gangs, brutal new "sports" like ultimate fighting, gory video games and violent language. Count how many synonyms there are for the word "kill." Now count how many there are for the word "love." Quite a difference. As we get more used to a hectic, impersonal lifestyle, malignatech is hard at work desensitizing us to brutality in all its many disguises. American culture has become one big reality show. Don't like your life? Hey, just change channels.

3) Spiritual—regardless of your religion or lack of it, all of us have some spiritual values. Hope, love, compassion are universal to the human spirit. Anything which seriously erodes those vital values is a major threat to our survival as genuine human beings. All of technology's colorful and noisy gadgets cannot mask its major contributions to our last fifty years: threats of nuclear and biological warfare; toxic, long-term

pollution everywhere; deadly, addictive, expensive drugs; radiation-spewing devices in hospitals (x ray machines and CAT scanners), businesses (computers), and homes (television sets and microwave ovens); dozens of new degenerative diseases.

Yes, we Americans are living longer lives in spite of all these new diseases, but what has happened to the quality of our lives? Statistics here are sobering. America ranks low on most quality of life charts:

- Life Expectancy—we must be number one or two, right? Not even close! The U.S. ranks 42nd. 2
- Health System—we're right after Costa Rica, at number 37. 3
- Maternal Health—women dying in childbirth can't happen in America, can it? Sadly, yes. We rank 18th in this category. 4
- Survival Rate for Newboms—we are tied for 32nd!

Interestingly, America also ranks low on overall happiness. A British psychologist, Adrian White, compiled data from over 100 studies in the field of happiness research on 178 countries. White used the data to create the first "World map of happiness" in 2007. Happiness here is based on a subjective "state of well-being" relating to such areas as health, work, and relationships. White's research at the University of Leicester revealed that health was the key indicator of happiness, followed by wealth, then access to basic education. Denmark came out as the happiest nation. America ranks 23rd.

Why have peace and harmony become such outmoded concepts? Why have confrontation and violence become so common?

Perhaps the key question here is why has organized religion failed to confront malignatech? With the exception of the Catholic Church in areas such as cloning and artificial birth control, modern religions seem strangely silent when it comes to discussing technology's dark side. Have all of our gadgets become so embedded into our culture that they are beyond reproach?

4) Financial—thanks to the magic of computers and the internet and our increasingly intrusive federal government, your financial freedom and privacy are steadily being eroded. Our contrived wars on drugs and terrorism have made it "necessary" to violate our Constitution by restricting Americans' access to their own money. Try to withdraw more than $10,000 in cash from any of your bank accounts and see what happens.

Making credit card purchases is becoming less secure all the time, as I noted in the computer chapter. Your financial records are being sold to anyone with the money to buy them. Even your access to obtaining credit is controlled by those all-powerful numbers known as credit scores. What's your Social Security number; what's your credit score; what's your cholesterol count; what's your IQ? Why are Americans so willing to live by numbers other people impose on them?!

The tragic irony of our nation's financial system is that the bewildering variety of investment and savings vehicles *all*—with the exception of precious metals—share one fatal weakness: they're all denominated in U.S. dollars. This simply means no matter how well these investments do, our rapidly weakening dollar is destroying your wealth. As we're already seeing in areas such as gas and food, inflation is heating up. It will eventually destroy the dollar and our economy. When our long-awaited Depression arrives soon, most Americans will be forced to look up from all those glowing screens and confront a bleak real world. Can we handle the shock of being yanked back to reality?

Solutions

How do we stop malignatech from destroying us? There are a number of simple steps we can all take to help bring technology under control:

1) ADMIT THE PROBLEM! If this book hasn't convinced you by now that malignatech is a real and growing problem, you are probably in denial that any of man's clever devices could ever do any harm. You need to reread malaxiom #2: "Technology was primarily developed to allow man to destroy nature for profit." If man is the smartest creature on Earth, why is he the *only* one to deliberately destroy his environment?

2) RETURN TO NATURE Once we admit that technology has literally been killing our Mother, we can begin the healing process, instead of hiding indoors behind those glowing screens, we need to get outdoors, feel the warmth of our nurturing sun and drink in the natural beauty surrounding us. Nature is not some primitive force in need of conquering. Nature is our family, and man is but one of its many children.

The most pragmatic way to celebrate Nature is to support a massive national campaign to develop, encourage, and use all forms of alternative energy. New developments in solar energy, such as thin-film solar panels, are finally pushing solar into the spotlight. Pioneer companies like First Solar and Nanosolar could be major energy producers in a few years, as they make solar energy prices competitive with fossil fuels. BCC Research, a technology market specialist, predicts the global solar energy market will expand from today's $13 billion to $32 billion by 2012. 6

3) CONFRONT CORPORATE AMERICA!! We've all heard the platitude that a corporation's primary responsibility is to maximize its profits for the good of its shareholders. Hogwash! A corporation's primary responsibility is not to endanger the health or safety of its shareholders—or of any other Americans!! In their mad scramble for profits, many U.S. companies have sold their souls to malignatech. As a shareholder of Mother Nature, your task is to recapture those lost souls before they do any more destruction. This all may seem like a truly impossible mission, but it's not. I've started the ball rolling for you by giving this dark side of technology a name and writing a book about it.

Controlling the damage corporate greed has been inflicting on America should be a relatively simple matter of enacting *very* strict new laws against pollution, deadly new drugs, etc. Unfortunately, such a route will be very slow, mainly because our bloated federal government has become so corrupted by corporate America that private citizens' voices rarely get heard.

Reforming this gigantic mess will have to start at the individual level. Join watchdog and conservation groups which aren't afraid to take governmental agencies to court. Vote with your purse or wallet by not buying violent video games or movies. Don't watch violent television shows or so-called sports like ultimate fighting. Tell everyone by your actions and words you consider violence something for losers.

Importantly, bail out of our babble culture and recognize it for the vicious cycle it actually is. When your federal government uses malignatech to start disastrous wars to further deny us our constitutional rights, then inflates our currency to destroy the economy, then allows pollution and toxic drugs to destroy our health, the American population gets very stressed out and begins to act like sheep. We herd together and seek solace in babbling to each other through our cell phones, text messages, emails, etc. While we're wasting time and money on these many distraction gadgets, our problems only get worse. The pathetic result is a fat, stupid, ignorant, spineless society which is a disgrace to our heritage. When our insanely indebted economy collapses soon, malignatech will have won a major victory. We will emerge from the ashes of a fiat currency leaner and poorer, but will we be any wiser?

Where Could Malignatech Lead Us

Let's take a look at where technology's dark side could take America if we don't change course soon:

Urged on by corporate ads and government stimulus checks, the American people have spent themselves into oblivion. Bankruptcy court has become the busiest place in the country. There are so many me mansions littering the landscape that homeless families are starting to take them over in return for maintenance work. Large banks are failing weekly.

Americans' physical and mental health continue their long decline thanks to our constant air, water, and ground pollution, garbage diet, and addiction to toxic drugs. A sharp increase in brain tumors among cell phone users leads to a long-overdue evaluation of their safety. The country is shocked to learn cell phones and especially their relay towers emit very toxic radiation *and* our friendly cell phone companies somehow forgot to warn us of this serious health hazard. As genetically modified foods take over most of our food supply, more delayed tests reveal long-term consumption of these foods to be dangerous to our genes, as in making us ill and even killing us. Since there are almost no normal seeds left anywhere but in a secret Norwegian ice cave repository, we are left with a maligna-choice: keep eating and getting sick or starving to death. Monsanto plants all over the U.S. are attacked by angry mobs.

Our federal government has invented a few more "wars" for us to fight. The first casualties in these battles are, of course, our constitutional rights. Americans are now so desperate for work they're doing most of the menial jobs illegal aliens used to do. Even with the immigration problem largely solved, the feds are polishing up the prototype of a tiny chip which can be imbedded in the wrist. With terrorist attacks and natural disasters becoming so frequent, the chip will supposedly allow emergency workers to learn your name, address, and vital medical data very readily. And, by the way, it will also list all your national identity numbers. Christians are aghast.

With possible terrorists lurking in every dark alley, surveillance companies have been the big growth market in our dormant economy. Snooping devices are now mandatory on all phones and computers. All new vehicles now come equipped—at no additional charge—with full-time GPS units, supposedly so police can locate your wheels if they are stolen. Vehicle speed recorders are being phased in.

Thanks to the oil companies' continued control of the federal government, peak oil has become an undisputed fact. Gas nationwide now averages over $8 a gallon and a new law makes it illegal to travel more than one mile alone in a four wheeled vehicle. Sales of dummies with moving heads and lips are soaring, as are sales of bicycles, scooters, and motorcycles.

The Supreme Court continues its on-going “reevaluation” of our Constitution. Because of a recent ruling, sales of new firearms have been outlawed. Liberal senators have introduced several bills to confiscate all firearms larger than .22 cal. pistols. The national militia movement, as a result, is gaining thousands of new members weekly, despite a new federal law outlawing all such groups. Many prominent essays, articles, and discussions in the media are emphasizing the need for more restrictions on individual rights to support all the various “war” efforts.

With the U.S. economic Depression now entering its third year, President Obama and the enlarged Federal Reserve System have been forced to confront our chaotic mess of an economy and 49 percent inflation rate with a partial return to a gold and silver backed currency. Since these two precious metals are now valued at $3,900 and $410 an ounce, respectively, speculators are eager to have the law finalized so they can trade in virtually worthless dollars for gold and silver coins. The feds have anticipated this move by making any attempts to convert dollars so enmeshed in red tape as to be almost impossible. Repeated requests by conservative senators and consumer groups for an audit of the gold in Fort Knox have been refused. One daring documentary filmer did manage to sneak a camera into our national gold depository and film giant stacks of lead bars painted gold. Now Americans know why Fort Knox hasn’t been audited since the 1950’s. That expose put a severe hold on the new currency plan. Meanwhile, our gigantic barter economy has reduced federal tax revenues to a fraction of their former levels, resulting in drastic cutbacks in all federal social aid programs. Destitute senior citizens are being encouraged to move in with their children, most of whom are furious with their parents for allowing once-supreme America to degenerate into a third world disaster.

You can elaborate on this grim scenario in your own imagination. Ask yourself if the quality of your life is better now than it was twenty, ten, or even five years ago. Your answer will probably surprise you.

Please reread my malaxioms, check out the books on my recommended reading list, and read your newspapers and magazines with a new, malignatech-oriented perspective. Better yet, start your own malignatech file with copies of relevant articles you find. You'll be amazed how quickly your file will fill up with scary articles. Remember that if we do not bring malignatech under strict control soon, this dark force will very likely destroy America.

I apologize for the fact that some of the data in this book will be obsolete by the time you read it. Technology's rate of change is getting faster and faster. I haven't delved into nanotechnology because the jury is still out on this radical new field of science. Revised editions of Malignatech should have more to say about nanotechnology.

I leave you with the prophetic words of one of the Renaissance's greatest writers. Sir Frances Bacon was an unusual combination of lawyer and philosopher who became a keen observer of Nature. He rejected the ancient method of relying on unproven assumptions. Instead, he pioneered the art of inductive reasoning, which led to our modern scientific method. Bacon believed that the only important knowledge for man was rooted in the natural world. Even though Frances Bacon died almost four hundred years ago in 1626, his brilliant mind seemed to anticipate the threat of malignatech when he wrote this saying:

"Nature, to be commanded, must be obeyed."

The End

FOOTNOTES

Chapter 1

1—*Reader's Digest,* March 2004, p. 20

Chapter 2

1 - Bruce McEwen, *The End of Stress As We Know It* (Washington, DC, 2002) p. 3

2 - American Heart Association website, June 12,2008

3 - Dr. David Williams, *Alternatives* newsletter, April 2003, p. 173

4 - Dr. David Williams, *Alternatives* newsletter, August 2002, pg's. 107—108

5 - Dr. David Williams, *Alternatives* newsletter, August 2002, p. 109

6 - OnLine Media, Inc. website, June 2008

7 - Dr. David Williams, *Alternatives* newsletter, January 2006

8 - Duff Wilson, *Fateful Harvest* (New York, 2001)

9 - *Life Extension Magazine,* September 2005, pg's. 66—68

10 - *Life Extension Magazine,* August 2005, p. 63

11 - *Life Extension Magazine,* August 2005, pg's. 65—67

12 - Gary Null, *The Complete Encyclopedia of Natural Healing* (Greenwich, CT, 2000) p. 87

13 - *Better Health Research* website, 2005

14 - Dr. David Williams, *Alternatives* newsletter, August 2005, pg's. 20—21

15 ~ Bruce McEwen, *The End of Stress As We Know It* (Washington, DC, 2002) pg's. 431—432

Chapter 3

1 - Jeffery Rothfeder, *every drop for sale* (New York, 2001) p. 8

2 - Jeffery Rothfeder, *every drop for sale* (New York, 2001) p. 21

3 - *World Hunger Year* website, 2007

4 - Jeffery Rothfeder, *every drop for sale* (New York, 2001) p. 5

5 - Steve Maxwell, *The Environmental Benchmarker and Strategist,* winter 2007

6 - *Emerging Trends Report* (Anchorage, March 2006)

7 - Theresa Watkins, *Gardening With Soul,* 2002

8 - *Emerging Trends Report* (Anchorage, March 2006)

9 -*U.S. International Information Programs* website, September 2004

10 - *U.S. International Information Programs* website, September 2004

11 - *Time Magazine* (New York, May 28,2007) p. 15

12 - Peter Dykstra, *CNN. com* website, August 1999

13 -*The Globe and Mail* website, July 31, 2000

14 - *U.S. International Information Programs,* September 2004

15 - Barnard R. Thompson, *Mexidatainfo* website, March 2005

16 - *Emerging Trends Report* (Anchorage, March 2006)

17 - *Water Partners International* website, June 2007

18 - *choike. org* website, July 2000

19 - *Corporate Research Project* website, May 2002

20 - *Emerging Trends Report (*Anchorage, March 2006)

21 - Carolyn Dean, *The Magnesium Miracle (*New York, 2007)

22 - idealist.org website, *Rehydration Project,* April, 2008

23 - *Time* Magazine, August 6, 2007

24 —*The Globe and Mail* website, July 31,2000

25 -Peter Dystra, *CNN.com* website, August 12, 1999

26 - *Los Angeles Times,* July 2007

27 - *Associated Press,* July 2007

28 - *Los Angeles Times,* July 2007

29 - *Philippine Sun Star,* December 13, 2004

30 - *Corporate Research Project* E-letter no. 24, May 2002

31 - Greg Palast, *The Progress Report*

32 - Bill Marsden, *The Center for Public Integrity,* February 2003

33 - Maude Barlow, *Blue Gold,* 2001

Chapter 4

1 - *Web MD,* October 10, 2007

2 - *MAF Communications,* Wellington, New Zealand

3 - *The Progress Report,* October 2007

4 - *Institute for Responsible Technology,* October 2007

5 - *Mothers for Natural Law,* September 27, 2007

6-Dr. Mercola website, Issue 1008, September 27, 2007

7 - *Foodreference. com* website, September 2007

8 - *Dr. Mercola* website, Issue 1008, September 27, 2007

9 - *Dr. Mercola* website, September 20, 2007

10 -*Dr. Mercola* website, September 20, 2007

11 - *Institute for Responsible Technology,* September 2007

12 - *Dr. Mercola* website, Issue 1008, September 27, 2007

13 —*Nutritional Well-Being AtoZ* website, October 2007

14- Dr. Mercola website, September 20, 2007

15- Jonathan Campbell, *50 Harmful Effects of Genetically Modified Foods,* September 27, 2007

16- *Foodreference.com* website, September 2007

17-*Foodreference.com* website, September 2007

18- Jonathan Campbell, *50 Harmful Effects of Genetically Modified Foods,* September 27,2007

Chapter 5

1 - *Human Genome Program* website, August 2006

2 - *Human Genome Program* website, August 2006

3 - Chris Wanjek, *Live Science,* January 2007

4 - *National Conference of State Legislatures,* April 2006

5 - *Human Cloning Foundation* website, 2002

6 - *Time* Magazine, February 4, 2008, p. 44

7 - Dr. Patrick Dixon, *Future Trends,* June 2007

8 -*Americans To Ban Cloning* website, 2002

9 -*American Bioethics Advisory Commission,* February 1987

Chapter 6

1 - Ronald Kotulak, *Inside The Brain,* Kansas City, 1996

2 - Dr. David Williams, *Alternatives* newsletter, April 2008

3 - Drake Bennett, *The Boston Globe,* January 13, 2008

4 - George F. Will, "Our education is worse than ever", April 24, 2008

5 - Ronald Kotulak, *Inside The Brain,* Kansas City, 1996

6 —*Pew Center On The States* website, February 28, 2008

7 - Ronald Kotulak, *Inside The Brain,* Kansas City, 1996

8 - Michael Ozner, *Life Extension Magazine,* March, 2008, pg's. 52—58

Chapter 7

1 - *Life Extension Magazine,* September 2005, p. 7

2 - *Bottom Line Personal* newsletter, December 15, 2005, p. 9

3 - Center for Disease Control website, MMWR, June 1996

4 - *Association of State and Territorial Directory of Health Promotion and Public Health Promotion* website

5 -*Angel Fire* website, February 2006

6 - Dr. David Williams, *Alternatives* newsletter, 2000, pg's. 1-3

7 - Dr. David Williams, *Alternatives* newsletter, 2000

8 - *The Wall Street Journal,* March 14, 2006, p. B4

9 - Dr. David Williams, *Alternatives* newsletter, September, 2002, p. 114

10 - Dr. Allan Magaziner, *New Hope Against Heart Disease*

11 - Dr. Julian Whitaker, *Health and Healing* newsletter

12 - *Time* magazine, November 6,2000, p. 103

13 - Dr. Julian Whitaker, *Health and Healing* newsletter

14 - The Independent Institute, FDAReview.org

15 - *San Francisco Chronicle,* November 23, 2004

16 - Mark Gold, *Aspartame...the BAD NEWS!*

17 -*Life Extension Magazine,* March 2006, pg's. 73—79

18 -*AARP Bulletin,* February 2006, pg's. 12 & 14

19 —*Medscape* from WEBMD

20 - *The Wall Street Journal,* March 19, 2006

21 - *South Florida Sun-Sentinel,* March 19, 2006

22 - *Life Extension Magazine,* March 2006, pg's. 42—45

23 - *The Wall Street Journal,* March 10,2006, p. A3

24 - Dale Steinreich, *Playing God at the FDA,* May 2,2005

25 - *Life Extension Magazine,* July 2008, pg's. 7-10

Chapter 8

1 - *Wikipedia* website

2 - *The Portland Phoenix,* February 27, 2004

3 - *The Boston Globe,* July 28, 2005, p. 6

4 - Robert F. Kennedy, Jr., *Rolling Stone Magazine,* June 20, 2005

5 - *Time Magazine,* January 9,2006, p. A4

6 - *Encyclopedia of Phytonutrient Treatments*

7 - *ETR Associates* website, April 2006

8 - Lisa Turner, *Do It Now Foundation,* August 2000

9 - *Shirley's Wellness Cafe* website, April 2006

10 - *National Institute on Drug Abuse* website, May 2006

11 - *National Drug and Safety League,* May 2006

Chapter 9

1—*Time Magazine,* January 28, 2008, p. 23

Chapter 10

1 - *The Wall Street Journal,* "Worried About Identity Theft," July 22, 2006, p. Al 1

2 - *Time Magazine,* March 20, 2006, p. 122

3 - *The Wall Street Journal,* "Worried About Identity Theft," July 22, 2006, p. Al 1

4 - Tom Herman, *The Wall Street Journal,* "Identity

Thieves Target Tax Refunds," March 12, 2008

5 - Jane J. Kim, *The Wall Street Journal,* "Credit Scorers Find New Ways To Judge You," March 11, 2008

Chapter 11

1 -Anne Galperin, *Nuclear Energy/Nuclear Waste, (*New York, 1992), p. 22

2 - Anne Galperin, *Nuclear Energy/Nuclear Waste, (*New York, 1992), p. 22

3 - *The Wall Street Journal,* December 14, 2005, p. Al 5

4 - *The Wall Street Journal,* December 14, 2005, p. A15

5 - Gail Kay Haines, *The Great Nuclear Power Debate, (*New York, 1985), pg's. 121 & 122

6 - Gail Kay Haines, *The Great Nuclear Power Debate, (*New York, 1985), pg's. 83 & 84

7 - Jay Taylor, *Energy and Energy Tech Stocks, (*New York, 2006), p. 5

8 - Gail Kay Haines, *The Great Nuclear Power Debate, (*New York, 1985), p. 86

9 - Gail Kay Haines, *The Great Nuclear Power Debate, (*New York, 1985), p. 92

10 - Michael Totty, *The Wall Street Journal,* June 30, 2008

11 - Anne Galperin, *Nuclear Energy/Nuclear Waste, (*New York, 1992), p. 22

12 - Jay Taylor, *Energy and Energy Tech Stocks, (Now* York, 2006), pg's. 5 & 6

13 - John G. Fuller, *We Almost Lost Detroit, (*New York, 1975), p. 25

14 - *The New York Times,* February 14, 2006

15 - *Time Magazine,* June 2, 2005

16 —John G. Fuller, *We Almost Lost Detroit, (*New York, 1975)

17 - Anne L. Galperin, *Nuclear Energy/Nuclear Waste,* (New York, 1992)

18 -*The Boston Globe,* October 20, 2005, p. A15

19 - John G. Fuller, *We Almost Lost Detroit,* (New York, 1975), p. 212

20 - John G. Fuller, *We Almost Lost Detroit,* (New York, 1975),p.233

21 - John G. Fuller, *We Almost Lost Detroit,* (New York, 1975)

22 - John G. Fuller, *We Almost Lost Detroit,* (New York, 1975), p. 241

23 - *Asbury Park Press,* February 12, 2006

24 - Gail Kay Haines, *The Great Nuclear Power Debate,* (New York, 1985), p. 44

25 - Gail Kay Haines, *The Great Nuclear Power Debate,* (New York, 1985), p. 51

26 - *Chicago Tribune,* February 21, 2006

27 - *Wikipedia*

28 -*Associated Press,* February 15, 2006

Chapter 12

1 - T. A. Heppenheimer, *Countdown,* (New York, 1997)

2 - Joseph J. Trento, *Prescription For Disaster,* (New York, 1987)

3 - Michael Cabbage and William Harwood, *Comm Check,* (New York, 2004)

4 - *The Boston Globe,* January 20, 2006

5 - *The Wall Street Journal,* October 5, 2005

Chapter 13

1 - Patrick Gorman, *Resource Consultants Inc.* website,

September 2007

2 - Linda Bilmes, *"Another year, another 300 billion," The Boston Globe,* March 16, 2008

3 —Patrick Gorman, *Resource Consultants Inc.* website, September 2007

4 - Eric Adams, *Popular Science Magazine,* "The Flying Laser Gun," April 2008

5 - Dr. Mercola website, August 2008

6 - Everest E. Riccioni, The Flawed V-22 Osprey and the Marine Corps, March 2006

Chapter 14

1 - Dr. David Williams, *Alternatives* website, April 8, 2008

2 - *Associated Press* website

3 - *WHO World Health Report* website

4 - *Project Censored website,* May 28,2008

5 - Dr. Mercola website, April 2008

6 - Bryan Walsh, *Time Magazine,* June 23, 2008, pg's. 62 & 63

Recommended Reading List

MEDICINE

Dr. Mercola's Website: Mercola.com

The Great Cholesterol Lie: The Truth About What Really Causes Heart Disease and How To

Avoid It by Malcolm Kendrick, M.D.

The Great Cholesterol Con by Anthony Colpo

Internal Bleeding by Robert M. Wachter, M.D. and Kaveh G. Shojania, M.D.

Our Daily Meds by Melody Petersen

Unstuck: Your Guide to the Seven-Stage Journey Out of Depression by James S. Gordon, M.D.

NUTRITION

Sweet Deception by Dr. Joseph Mercola

Seeds of Deception by Jeffrey Smith

The Whole Soy Story: The Dark Side of America's Favorite Health Food by Dr. Kaayla Daniel

Life Extension Magazine—website: wwwlef.org

The Sugar Fix by Richard Johnson

ENVIRONMENT

Every Drop For Sale by Jeffrey Rothfeder

High Tide by Mark Lynas

Our Stolen Future by Theo Colburn, Dianne Dumanoski, and John Peterson Myers

Fateful Harvest by Duff Wilson

The Weather Makers by Tim Flannery

Plague Wars by Tom Mangold and Jeff Goldberg

Contaminated Communites by Michael R. Edelstein

Our Final Hour by Martin Rees

Terra by Michael Novacek (how man is causing mass extinction of other species)

No Place To Hide by Robert O'Harrow, Jr.

Last Child in the Woods: Saving Our Children from Nature Deficit Disorder by Richard Louv

Bottlemania by Elizabeth Royte

Made in the USA
Monee, IL
28 February 2026

45151169R00095